A Manager's Guide to Green-Lean-Six Sigma Application

Amna Farrukh
Newcastle Business School
Northumbria University
Newcastle, UK

Aymen Sajjad
School of Management and Marketing
Massey Business School
Massey University
Auckland, New Zealand

ISBN 978-3-032-21110-1 ISBN 978-3-032-21111-8 (eBook)
https://doi.org/10.1007/978-3-032-21111-8

This Springer imprint is published by the registered company Springer Nature Switzerland AG
The registered company address is: Gewerbestrasse 11, 6330 Cham, Switzerland

Amna Farrukh · Aymen Sajjad

A Manager's Guide to Green-Lean-Six Sigma Application

Tools and Practices

Springer

To our Families, in Recognition of their Relentless Support, Love, Perseverance, and Unwavering Belief in us throughout this Endeavour.

Competing Interests The authors have no competing interests to declare that are relevant to the content of this manuscript.

Contents

About the Authors

Amna Farrukh is an Assistant Professor of Operations and Supply Chain Management at Northumbria University, Newcastle, United Kingdom. She holds a Ph.D. in Engineering Management from Massey University, Auckland, New Zealand. She also holds a Six Sigma Black Belt certification. Her areas of interest are sustainable operations and supply chain management, supply chain resilience, digitalization, the circular economy, green lean six sigma, environmental sustainability, and waste management.

Aymen Sajjad is an Associate Professor of Sustainable Business Management and Discipline Lead for Management and International Business at the School of Management and Marketing, Massey Business School, Massey University. He holds a Ph.D. in Sustainable Supply Chain Management from Massey University, New Zealand. Aymen's research focuses on sustainability issues that confront organizations and societies. He has received several research awards and was recognized in 2025 among the World's Top 2% Scientists list—a prestigious list published annually by Stanford University and Elsevier. Aymen is an Associate Editor of Business Strategy and the Environment, Corporate Governance: The International Journal of Business in Society, Journal of Management & Organization, FIIB Business Review Journal, and the International Journal of Sustainable Society. Aymen is also an Editorial Advisory Board Member for the Review of International Business and Strategy, Management and Sustainability: An Arab Review, and an Editorial Review Board Member for the Journal of Global Responsibility. He has published in top-tier business management and sustainability journals (e.g., Journal of Economic Behavior and Organization, International Journal of Management Reviews, Business Strategy and the Environment, International Journal of Operations and Production Management, and Journal of Social Indicators Research). His research interests include Sustainable Production and Operations Management, Sustainable Supply Chain Management, Business Sustainability, Corporate Social Responsibility, Sustainability Leadership, and the United Nations Sustainable Development Goals (SDGs). He is also the Deputy Director (Research) of the Sustainable Futures Research Hub at Massey Business School.

About the Authors

Abbreviations

AI	Artificial intelligence
ANOVA	Analysis of variance
CBA	Cost-benefit analysis
CE	Circular economy
CLSS 4.0	Circular Lean Six Sigma 4.0
CO	Carbon monoxide
CO_2	Carbon dioxide
CPS	Cyber physical system
CPS	Cyber-physical system
CSFs	Critical success factors
DFE	Design for the environment
DFSS	Design for six sigma
DFSS	Design for six sigma
DMAIC	Define, measure, analyze, improve, control
DOE	Design of experiment
ECRS	Eliminate-combine-rearrange-simplify
EMS	Environmental management system
EnMS	Energy management system
ENP	Electroless nickel plating
EPA	Environmental Protection Agency
E-VSM	Environmental value stream mapping
FMEA	Failure mode and effect analysis
GHG	Greenhouse gases
GLSS	Green-Lean-Six Sigma
ICBV	Intellectual capital-based view
ICT	Information communication technologies
Industry 4.0	I4.0
IoT	Internet of Things
IPCC	Intergovernmental Panel on Climate Control
ITBV	Institutional theory-based view
JIT	Just-in-time

LCA	Life cycle analysis
LSS	Lean six sigma
MCDM	Multi-criteria Decision-Making
MICMAC	Matrice d'Impacts Croisés Multiplication Appliqués à un Classement
NO_x	Nitrogen oxide
NRBV	Natural resource-based view
OECD	Organisation for Economic Co-operation and Development
PCA	Process capability analysis
PDCA	Plan-do-check-act
QFD	Quality function deployment
RFID	Radio frequency identification
SAP	System Analysis and Program Development
SIPOC	Supplier-input-process-output-control
SLSS	Sustainable lean six sigma
SMED	Single-minute exchange of die
SOPs	Standard operating procedures
SO_x	Sulfur oxide
SPC	Statistical process control
TBL	Triple bottom line
TPM	Total productive maintenance
TQM	Total quality management
VOC	Voice of the customer
VOCs	Volatile organic compounds
VSM	Value stream mapping
WWF	World Wildlife Fund
3R	Reducing reusing recycling
5S	Seiri, seiton, seiso, seiketsu, shitsuke

List of Tables

Chapter 1
Introduction to Green-Lean-Six Sigma Approach

1.1 Introduction

In recent times, critical environmental issues, including climate change, natural resource depletion, habitat destruction, pollution, ocean acidification, loss of biodiversity, deforestation, and waste generation, have propelled firms, governments, and society to embrace environmental sustainability and foster natural resource conservation (Esty and Winston 2009; Hariyani and Mishra 2022a; Yadav et al. 2023a, b). As such, the excessive consumption of natural resources, including water, minerals, fossil fuels, fertile soil, and forests, led to a myriad of complex environmental issues and a serious risk of ecological imbalance for both developed and developing economies (Danta and Rath 2024; Hasimoglu 2021; Lu and Wang 2023). Further, growing natural resource depletion affects biodiversity, disrupts natural cycles, causes greenhouse gas (GHG) emissions, and exacerbates climate change (Lu and Wang 2023). According to the United Nations Development Programme (2020), natural resource consumption has increased to 100.6 million, a fourfold increase from 1970, with an average annual growth rate of 5.5% (Danta and Rath 2024).

In response to complex and increasingly growing sustainability concerns, the World Commission on Environment and Development (WCED), also known as the Brundtland Commission, proposed the sustainable development concept, which refers to "the [inclusive] development that meets the needs of the present without compromising the ability of future generations to meet their own needs" (WCED 1987, p. 43). In a corporate environment, sustainable development (also referred to as 'sustainability' in a business context) is broadly conceived as a holistic concept that involves incorporating social, economic, and environmental concerns into strategic and operational decision-making to simultaneously achieve both short-term and long-term business performance. As Székely and Knirsch (2005) expound, sustainability from a business perspective involves:

A. Farrukh and A. Sajjad, *A Manager's Guide to Green-Lean-Six Sigma Application*,
https://doi.org/10.1007/978-3-032-21111-8_1

> Sustaining and expanding economic growth, shareholder value, prestige, corporate reputation, customer relationships, and the quality of products and services as well as adopting and pursuing ethical business practices, creating sustainable jobs, building value for all the stakeholders and attending to the needs of the underserved. (p. 628)

Some scholars, however, have focused more on the environmental dimension while explicating the role of the corporate sector in promoting sustainable development as "the ability of a firm to nurture and support growth over time by effectively meeting the expectations of diverse stakeholders" (Neubaum and Zahra 2006, p.121).

As one of the primary users of pristine natural resources, manufacturing firms continue to face a multitude of environmental challenges due to the excessive use of natural resources and energy consumption requirements in industrial processes that cause GHG emissions, harmful air and water pollution, and other forms of hazardous waste and damaging substances (e.g., Esty and Winston 2009; Mohamad et al. 2022; Robinson 2009). For example, the chemical industry faces critical sustainability challenges, including natural resource depletion, the use of hazardous materials, and employees' and workplace safety issues (Liew et al. 2014; Mohan et al. 2021; Papafloratos et al. 2023). Similarly, the textile industry is considered one of the largest water-consuming industries and is often criticized for environmental pollution, including air emissions, wastewater, energy consumption, and discharge of hazardous chemicals (Luoma et al. 2022). In addition, the leather industry faces significant environmental sustainability challenges, including wastewater generation, human health and safety issues, and the use of toxic chemicals (Brun and Cicculllo 2022; Chen et al. 2023; Omoloso et al. 2021; Sultana et al. 2023). Moreover, the pharmaceutical industry faces numerous sustainability challenges, including excessive energy consumption, GHG emissions, hazardous chemicals, and waste generation (Khan and Ali 2022; Kharat et al. 2025; Tsolakis et al. 2019).

Emerging challenges, including economic concerns, growing consumer demand for environmentally friendly products, stringent environmental legislation and policies, and market dynamics, have also influenced the organizations to adopt sustainable practices (Yadav et al. 2023a, b). Rapid industrial development and human-led activities have increased global warming and GHG emissions. According to the Intergovernmental Panel on Climate Change (IPCC) report, global warming has reached +1.1 °C compared with the pre-industrial era, jeopardizing the goal of limiting global warming to 1.5 °C, set by the Paris Agreement. Further, global warming could reach +3 °C by 2100, exceeding the safety threshold limit set by the scientific community if immediate actions are not taken (IPCC 2024). The report also emphasized extreme weather conditions, rising sea levels, global food security risks, and loss of biodiversity as contemporary sustainability issues calling for immediate attention and s from stakeholders.

One of the significant contributors to GHG emissions is the industrial sector (e.g., Panagiotopoulou et al. 2022), which makes up around 35% of the total emissions. The reasons are associated with the conventional methods of manufacturing products used by businesses in various countries (Wang et al. 2019). Further, the increasing GHG emissions have adverse environmental and social sustainability impacts, resulting in severe health issues (Kwon 2020). Accordingly, manufacturers continue to face increasing pressure to implement sustainable strategies in their operations in response to growing environmental issues and stringent environmental regulations (Yadav et al. 2023a, b). In response, firms heavily invest in their research and development activities to adopt sustainable manufacturing practices (Pandey et al. 2018; Sharma et al. 2017; Yadav et al. 2023a, b).

Pro-environmental strategies, including green manufacturing, lean manufacturing, sustainable manufacturing, Six Sigma, and agile manufacturing, are gaining traction among manufacturing firms aiming to achieve operational efficiencies and environmental goals (Ishak et al. 2024; Mohan et al. 2024). In particular, green manufacturing, lean manufacturing, and Six Sigma are increasingly employed to optimize industrial operations and environmental performance (Yadav et al. 2023a, b). The following sections describe these selected strategies and their interrelationships.

1.2 Sustainable Performance

Sustainable performance refers to simultaneously enhancing economic, environmental, and social performance in internal business operations and the broader societal contexts. Firms aim to achieve sustainable performance through the triple bottom line (TBL) objectives comprising the economic, environmental, and social performance indicators, while maintaining a balance between these aspects to produce optimal performance (Elkington 1998; Julie et al. 2024; Shou et al. 2019; Siefan et al. 2025). Different indicators are used to assess an organization's sustainable performance. Economic performance indicators involve cost efficiency; lead time reduction; on-time delivery; and increased profitability, productivity, efficiency, and customer satisfaction (Faulks et al. 2021; Mishra et al. 2023; Rahman et al. 2023; Shou et al. 2019). Conversely, environmental performance indicators entail resource efficiency; waste, emission, and pollution reduction; raw material, water, and energy conservation; recycling; use of renewable and sustainable materials; and avoidance of hazardous materials and chemicals (Mu et al. 2023; Nutsugah et al. 2021; Owusu-Manu et al. 2023; Turker and Altuntas 2014). Social performance indicators include employee satisfaction, wage equity, gender inclusivity, workplace safety, employee training and education, community development, and social welfare (Cao et al. 2023; Farrukh et al. 2025; Ruel and Fritz 2021; Silva et al. 2023).

1.3 Foundation of Green-Lean-Six Sigma: Underpinnings and Context

1.3.1 Green Strategy

Green strategy focuses on improving organizations' environmental performance and achieving operational and financial objectives, as organizations are becoming more conscious about their manufacturing operations due to the increasing environmental impacts and resource depletion issues (Appiah et al. 2025; Garza-Reyes 2015; Yu and Jin 2022). Green strategy focuses on determining, measuring, evaluating, and controlling environmental waste by addressing product- and process-related issues (Sadiku et al. 2023; Sohani et al. 2022; Yadav et al. 2023a, b). Further, green strategy deals with environmental protection and conservation of resources while reducing the negative environmental impacts of organizations resulting from the manufacturing of products (Bapat et al. 2024; Deif 2011; Hariyani and Mishra 2022a). Incorporating green strategy in business processes and operations helps companies comply with environmental regulations and produce environmentally friendly products, which in turn enables them to achieve a sustained competitive advantage (Bapat et al. 2024; Chiou et al. 2011; Salifu et al. 2025). Other benefits of adopting a green strategy include improving the firm's financial, market, and environmental performance through enhanced corporate image and resource efficiency (Chotia et al. 2023; Hariyani and Mishra 2022a; Rashid et al. 2025). Additionally, green strategy enables firms to reduce the environmental cost of penalties, fines, and disposal of hazardous waste, including chemicals, pollutants, and GHG emissions (Chiou et al. 2011; Rusinko 2007; Shang et al. 2010).

Green manufacturing focuses on creating products through environmentally friendly processes, conserving resources, minimizing waste, and considering the safety concerns related to employees, consumers, and the community. Green or cleaner manufacturing is "a systematically organized approach to production activities, which has positive effects on the environment. These activities encompass resource use minimization, improve eco-efficiency and source reduction, in order to improve the environmental protection and to reduce risks to living organisms" (Glavič and Lukman 2007, p. 1879). In addition, the notion emphasizes using sustainable practices in manufacturing products, such as cleaner production methods and eco-efficiency (Amjad et al. 2021; Faulkner and Badurdeen 2014; Lu et al. 2022). Overall, green manufacturing includes the production of goods in a sustainable manner, emphasizes the total product life cycle, and covers the economic, environmental, and social implications (Elkington 1998). Taking a total product life cycle perspective is imperative to initiate sustainable, innovative practices to gain a competitive advantage (Ijaz et al. 2024; Xing and Liu 2023; Yadav et al. 2023a, b). Owing to the increasing energy consumption, waste generation, air emissions, and contamination, governments in developed and developing economies compel organizations to adopt sustainable practices by introducing stringent laws, frameworks, and regulations.

Green strategy stresses the need for waste and pollution reduction, and natural resource conservation. In addition, many organizations have included the "Go Green" agenda in their mission statement to eliminate environmental waste (Gaikwad and Sunnapwar 2020; Yadav et al. 2023a, b). In this context, various green tools are utilized for emission and waste reduction, including 3R, life cycle analysis, eco-design, environmental management systems (EMS), eco-labelling, and design for the environment (DFE) (de Souza Costa et al. 2021; Yadav et al. 2023a, b). Green strategy also includes practices such as green purchasing, green design, and green packaging (Khan et al. 2023; Mahmoud et al. 2022; Rasheed et al. 2025; Yadav et al. 2023a, b).

1.3.2 Lean Strategy

The lean paradigm stresses the importance of 'lean manufacturing' or 'lean production', which is used to improve the manufacturing operations by reducing waste. Lean refers to the management philosophy focusing on activities that add value to the products and eliminating the unnecessary ones (Austin and Adebayo 2021; Chugani et al. 2017; Holweg 2007; Nishad et al. 2024). By identifying the non-value-added activities, lean manufacturing helps reduce waste and conserve resources, which in turn leads to cost reduction and quality improvement (Drohomeretski et al. 2014a, b; Mathiyazhagan et al. 2021; Narassima et al. 2025). The five principles of lean strategy (value, mapping the value stream, flow, pull, and continuous improvement) are directed at improving customer value (Chugani et al. 2017; Klein et al. 2024; Sony and Naik 2019). Principle 1 (value) focuses on understanding the customers' requirements and expectations (i.e., what is of value to the customer) (Middleton and Sutton 2020). It also emphasizes the internal and external factors affecting the customer decisions for creating value. Further, it identifies the value and non-value-added activities to indicate what the customer is willing to pay as part of the value creation in a product or service (Ahmed 2018).

Principle 2 (value stream mapping) activities emphasize the importance of the right product, as well as waste identification and waste elimination (Cuer et al. 2025; Gao et al. 2025). In this regard, activities are identified from the start to the finish of a product for achieving customer value (Helmold et al. 2022). Principle 3 (flow) focuses on a constant flow of products without interrupting the operations by avoiding bottlenecks in the process. Moreover, it emphasizes that buffers and large batches should be avoided or reduced (Jaskari 2014). Principle 4 (pull) refers to reducing overproduction, such as applying the just-in-time (JIT) strategy to reduce inventories (including raw materials), increase process efficiency, reduce operational waste, and ensure the flow of products from ordering to distribution (Ahmed 2018). JIT strategy aims to deliver the right product at the right time and in the right quantity. This principle also emphasizes not producing or delivering anything 'upstream' unless it is needed 'downstream' (Ahmed 2018, p. 183). Principle 5 (perfection) focuses on successfully implementing all four previous principles to achieve better operational performance continually, leading to perfection. It aims to achieve continuous improvement while delivering products and services according to customer requirements at an optimal price (Ahmed 2018). Continuous improvement depends on the value addition and flexible and continuous flow (Jaskari 2014). The overall aim is to align the production operations, considering the customer requirements, with minimum waste and operational cost (Jaskari 2014).

Lean manufacturing originated from Toyota Production Systems (TPS), which relates to waste reduction and removing non-value-added activities from industrial processes (Aadithya et al. 2023; Hariyani and Mishra 2022a; Sordan et al. 2025). TPS mainly emphasizes the eight types of operational waste, including overproduction, transportation, unnecessary motion, excessive inventory, defects, overprocessing, waiting, and underutilized human resources (Hariyani and Mishra 2022a; Womack et al. 2007). To remove these wastes from the processes and supply chain, firms apply value stream mapping—a lean tool to differentiate between the non-value-added and value-added activities (Carrijo et al. 2024; Hariyani and Mishra 2022b; Shabeen and Krishnan 2022). In this context, firms focus on embedding the customers' voice in their processes to avoid operational waste (Jayanth et al. 2020).

Scholars have identified several benefits of lean manufacturing, such as reducing cycle time, lead time, work-in-progress, waiting time, inventory, machine downtime, and operational cost (Gebeyehu et al. 2022; Ghosh 2013; Gochel et al. 2022; Hofer et al. 2012; Joseph et al. 2021). This further includes enhancing customer satisfaction and improving process optimization; as well as increasing inventory turnover, product quality, labour productivity, and the firm's competitive advantage in the long run (Adeodu et al. 2021; Ghosh 2013; Karim and Arif-Uz-Zaman 2013; Pingyu and Yu 2010). The definition of value creation in lean manufacturing has also been extended by including the circular economy criteria of waste and emission reduction, resource conservation, and environmental safety. This suggests that value creation is not only done to fulfil customers' needs but also to utilize resources in a way that maximizes their usability at all stages of the life cycle, including end-of-life management of a product (Farrukh et al. 2023; Nadeem et al. 2019).

Prior studies have revealed the successful application of lean strategy not only in the manufacturing context but also in service industries such as hospitals and educational institutions. Lean strategy comprises several tools, including value stream mapping (VSM), single-minute exchange of die (SMED), 5S (seiri, seiton, seiso, seiketsu, and shitsuke), Kanban, Poka-yoke, etc. (Al-Rifai 2024; Chugani et al. 2017; Dahiya et al. 2022; Martins et al. 2023; Wong and Wong 2014). Moreover, various lean tools emphasize employee contributions to continuous improvement in organizations, such as Kaizen, 5S, Poka-yoke, Kanban, and VSM (Ciannella and Santos 2022; Simões et al. 2024; Singh et al. 2021; Slavina and Štefanić 2024).

1.3.3 Six Sigma Strategy

The Six Sigma strategy, pioneered by Motorola, Inc., focuses on reducing product and process variations by controlling defects through various tools that concentrate on producing 99.99966% defect-free products (Hariyani et al. 2023; Pande et al. 2000). Six Sigma assists in defect reduction and control of process variations (Condé et al. 2023; Sadikin 2023; Setiawan 2025). Six Sigma strategy refers to organizations' breakthrough and continuous improvement practices to reduce operational costs, enhance customer satisfaction, and produce high-quality products (Yadav et al. 2023a, b). Six Sigma is recognized as a "controlled approach that minimizes variation in the process so that defects are minimized to less than 3.4/ M opportunities" (Yadav et al. 2023a, b, p. 764).

Six Sigma strategy comprises tools such as statistical process control (SPC), statistical quality control, cause-and-effect analysis, analysis of variance (ANOVA), regression analysis, and correlation analysis. Six Sigma tools are also utilized systematically under the two popular methodologies—DMAIC (define, measure, analyse, improve, and control) and DMADV (define, measure, analyse, design, and verify) (Bhat et al. 2020; Patel and Desai 2018; Sreedharan et al. 2018). The DMAIC methodology helps identify, analyse, improve, and control an organization's existing manufacturing processes and is considered a reactive approach to problem solving (Condé et al. 2023; Hilton and Sohal 2012). Conversely, the DMADV methodology, also known as design for Six Sigma, assists organizations in designing new products and processes for defect-free production (Belamkar 2016; Chugani et al. 2017). DMADV is recognized as a proactive approach to problem solving. Table 1.1 presents an overview of Six Sigma's DMAIC and DMADV methodologies.

A range of Six Sigma tools, including design of experiments (DOE), cost–benefit analysis, and statistical process control (SPC), help in reducing operational cost, increasing customer satisfaction, product quality, process efficiency, and robustness, and reducing cost due to robust process design (Antony 2006; Hariyani et al.

Table 1.1 Differences between DMAIC and DMADV

DMAIC	DMADV
D: Defining a process that requires development	D: Defining objectives align with customer requirements
M: Measuring critical factors affecting process efficiency	M: Measurement of factors affecting quality attributes
A: Analyzing relevant factors for process improvement	A: Analysis of alternative solutions
I: Process improvement by developing and executing a solution	D: Designing alternative method
C: Controlling process to ensure sustainable improvement	V: Verification of the method before implementation

Source Adapted from Chugani et al. (2017)

2023). Along with achieving the operational benefits, the Six Sigma methodology can address organizations' environmental sustainability targets (Chugani et al. 2017). Despite reducing the environmental issues, many organizations are unaware of the potential sustainability benefits of the Six Sigma methodology (Antony et al. 2019a, b; Sony et al. 2019).

Scholars have highlighted Six Sigma's environmental advantages concerning energy saving, resource conservation, waste reduction, and pollution prevention (Al-Senani et al. 2025; Chugani et al. 2017; Jin et al. 2024; Raduan et al. 2025; Tsarouhas and Sidiropoulou 2024). Six Sigma tools help organizations improve the shop floor performance, such as root cause analysis, which assists in identifying and controlling quality issues, as one of the key Six Sigma practices (Yadav et al. 2023a, b). The key objectives of the Six Sigma strategy include mitigating the leading causes of the problems and improving the product quality through process improvement (Kumar et al. 2013; Yadav et al. 2023a, b).

1.4 Evolution and Interlinkages Between Green, Lean, and Six Sigma Strategies

1.4.1 Green-Lean Strategy

The distinctions between lean and green strategies can be understood by their underlying philosophy, as the former focuses on non-value-added activities and operational waste, and the latter emphasizes environmental waste (Duarte and Cruz-Machado 2013). Despite their key difference, the United States (US) Environmental Protection Agency (EPA) (2007) highlighted the significance of an integrated Lean-Green strategy by emphasizing the interrelationship between operational and environmental waste (Table 1.2). According to the EPA, the lean Mudas (types of waste) contribute to environmental waste, for example, excessive inventory results in environmental waste in the form of excessive energy consumption in the warehouse and potential risk of disposal to the environment from unused and expired products. Similarly, inefficient facility layout could result in the excessive requirements of transportation of raw materials and work-in-progress inventory, which further leads to energy consumption and air emissions (Carvalho et al. 2011). Accordingly, scholars have argued that the execution of lean and green should be integrated to achieve optimum results (Garza-Reyes 2015; Ng et al. 2015; Sagnak and Kazancoglu 2016).

The interlinkages among these strategies are also apparent from the extant literature in overcoming each other's limitations by utilizing their relevant tools. In this regard, one study indicated the integration of different lean tools, such as value stream mapping (VSM) and life cycle assessment (LCA), to improve the environmental performance of an automotive parts component manufacturing company in India (Hoogmartens et al. 2014). The study highlighted that the LCA (a green tool) was used to classify the environmental impacts of different aspects, such as water

Table 1.2 Environmental impacts linked to manufacturing waste

Waste type	Environmental impact
Defects	• Raw materials are consumed to produce defective products • Defective components require recycling or disposal • Additional space is required for rework and repair, increasing energy use for heating, cooling, and lighting
Waiting	• Potential material spoilage or component damage, causing waste • Wasted energy from heating, cooling, and lighting during production downtime
Overproduction	• Raw materials are consumed to make products that are not in demand or cannot be consumed in due course • Extra products may spoil or become obsolete, requiring disposal
Movement and transportation	• More energy is used for transport • Emissions from transport • Additional space is required for work-in-progress (WIP) movement, which increases lighting, heating, and cooling demand and energy consumption • Additional packaging may be required to protect components during movement
Inventory	• More packaging to store WIP waste from deterioration or damage to stored WIP • Extra materials are needed to replace the damaged WIP • More energy is used to heat, cool, and light the inventory space
Complexity and overprocessing	• Additional parts and raw materials are consumed per unit of production • Unnecessary processing increases waste, energy use, and emissions
Unused creativity	• Fewer suggestions for pollution and waste minimization opportunities

Source EPA (2003)

consumption, carbon emissions, air acidification, and energy consumption. Then the associated effects were incorporated with VSM (a lean tool) to develop an overall picture of the environmental performance of the manufacturing firm (Hoogmartens et al. 2014). Similarly, another study simultaneously implemented VSM and LCA to identify and measure the environmental waste in the construction industry (Banawi and Bilec 2014). Although VSM can highlight various types of waste, it cannot measure environmental aspects. The LCA has filled this gap by making it easier to perform environmental assessments (Banawi and Bilec 2014); however, it could not highlight all possible wastes that have an impact on environmental performance, and this limitation was addressed by lean manufacturing through one of the key tools—VSM. In this way, lean and green complement each other. In a similar vein, Fecroq et al. (2016) integrated 3R (a green tool) and Mudas (lean 7 waste analysis) in a lean/

3R matrix to highlight the waste reduction and environmental impact of manufacturing organizations. Another study emphasized integrating lean and environmental management systems (EMS) (e.g., ISO 14001) based on their standard waste reduction features to improve operational and environmental performance (Ahmed and Mathrani 2019).

An integrated Green-Lean framework can also improve the firm's financial and ecological performance (Marco-Ferreira et al. 2020; Yadav et al. 2023a, b). According to Pampanelli et al. (2014), an integrated Green-Lean strategy can reduce resource consumption by 40–60% and minimize the energy-related costs for manufacturing firms. Several authors have emphasized the Green-Lean strategy for improving an organization's sustainability performance. For example, Garza-Reyes (2015) developed a Green-Lean framework based on a systematic literature review to simultaneously improve operational and environmental performance through waste reduction, emission control, and optimal resource utilization. Similarly, Lu et al. (2019) conducted a study in the construction sector and proposed an integrated Green-Lean framework to reduce GHG emissions. Another study developed a Green-Lean model for manufacturing organizations to analyse and improve their manufacturing operations and mitigate the barriers while integrating the two approaches (Yadav et al. 2021a, b, 2023a).

1.4.2 Lean Six Sigma Strategy

The term Lean Six Sigma was first introduced in the early 2000s. Several studies acknowledged the significant benefits of utilizing Lean Six Sigma as an integrated approach, overcoming each strategy's limitations (Chaudhary et al. 2023). Numerous researchers have investigated and highlighted the successful integration and deployment of Lean Six Sigma in various industrial sectors, including construction, dairy, healthcare, and food manufacturing (Fayyaz et al. 2025; Jamil et al.; Marrucci et al. 2020; Pepper and Spedding 2010; Sharma and Singh 2025; Yadav et al. 2023a, b).

Lean Six Sigma not only improves organizational performance in the manufacturing sector but also helps service organizations to achieve the desired results (Lizarelli et al. 2023; Raduan et al. 2025; Siefan et al. 2025; Yadav et al. 2021a, b). In this regard, a study presented a Lean Six Sigma framework in the banking sector for improving shareholder management and revealed the benefits of using the Lean Six Sigma strategy in the form of improved financial performance with a profit of Indian Rupee (INR) 1.6 million (Sunder 2016). Similarly, another study investigated the application of lean, Six Sigma, and Lean Six Sigma in manufacturing industries and highlighted the cost reduction as a significant reason for executing the integrated approach of Lean Six Sigma (Drohomeretski et al. 2014a, b).

Prior studies have focused on the premise that lean and Six Sigma strategies help manufacturing organizations achieve their environmental goals (Cherrafi et al. 2016). The lean and Six Sigma combination is recognized as an integrated strategy to improve the organization's operational performance (Garza-Reyes 2015). This integrated strategy shares common goals of lean and Six Sigma paradigms, such as waste reduction, quality improvement, process optimization, continuous improvement, and customer satisfaction (Farrukh et al. 2020; Maleyeff et al. 2012).

Both lean and Six Sigma are well-recognized strategies to improve organizations' operational performance (Antony et al. 2019a, b; Yadav et al. 2023a, b). Scholars have emphasized integrating lean and Six Sigma methodologies in different ways. For example, one study highlighted the systems approach to combine the two strategies (Chugani et al. 2017; Pojasek 2003). Similarly, another study stressed combining the Plan-Do-Check-Act cycle (PDCA) with the DMAIC technique to achieve better organizational performance through adequate planning, execution, and measurement of the organization's objectives (Burghall et al. 2014). In addition, both techniques (PDCA and DMAIC) follow a cyclic approach until the goals are achieved; hence, they are recognized as continuous improvement techniques.

In contrast, few studies have emphasized the execution of lean and Six Sigma methodologies separately, instead of in combination, as the combined implementation may not achieve the desired success (Salah et al. 2010). In this context, organizations first need to deploy the Six Sigma methodology to enhance the process performance and then apply the lean strategy to improve the system's overall efficiency (Crawford 2004). Conversely, Snee (2004) stressed the implementation of lean in the first stage to identify and reduce process waste, and then use the Six Sigma methodology. Most of the studies, however, emphasized the combined application of lean and Six Sigma to gain the optimum operational performance (Chugani et al. 2017; Salah et al. 2010).

1.4.3 Green-Lean-Six Sigma Strategy

A plethora of studies are available investigating the environmental impacts of green, lean, and Six Sigma strategies, both individually and in combination with each other, such as Lean-Green and Lean Six Sigma (Banawi and Bilec 2014; Cherrafi et al. 2016; Pampanelli et al. 2014). Nevertheless, the growing body of research emphasizes integrating all three green, lean, and Six Sigma strategies as a holistic approach to gaining sustainable operational, environmental, and social benefits.

While the lean strategy focuses on process improvement and creating value for the customers by removing non-value-added activities (Garza-Reyes 2015; Yadav et al. 2023a, b), it is unable to consider environmental waste and variation in processes. Hence, the above shortcomings of lean strategy can be tackled by the green and Six Sigma strategy that helps in producing environmentally friendly products, along with improving quality and reducing costs (Yadav et al. 2023a, b).

Green manufacturing focuses on conserving natural resources and the environment, minimizing the harmful impact of products and processes, and developing eco-friendly products through implementing sustainable practices (Huo et al. 2019; Yadav and Gahlot 2022; Yadav et al. 2023a, b). Although the green strategy addresses ecological issues, it does not consider the aspects of process variation in manufacturing operations. To overcome this limitation, Six Sigma helps the green strategy by reducing and controlling the variation in the processes, which could result in resource conservation (such as reducing defects) and environmental protection (Yadav et al. 2023a, b). Nevertheless, Six Sigma alone cannot address the environmental waste in organizations' processes (Farrukh et al. 2020; Sun et al. 2017).

Despite the operational advantages of Lean Six Sigma, this approach cannot address organizations' environmental and social performance. Accordingly, to overcome the limitations, scholars have emphasized integrating green strategy with Lean Six Sigma to obtain better environmental and social performance and operational benefits. As such, an integrated Green-Lean-Six Sigma (GLSS) strategy helps organizations in improving resource efficiency, reducing negative environmental impacts, and enhancing customer satisfaction (Chaudhary and Mudimela 2022; Yadav et al. 2023a, b). In recent years, organizations have realized the environmental benefits of maintaining operational performance using an integrated GLSS approach. Several authors have emphasized the integrated GLSS strategy as a way to overcome the limitations of each strategy (Farrukh 2025; Fayyaz et al. 2025; Javed 2025; Rathi et al. 2021).

1.5 Green-Lean-Six Sigma—An Integrated Approach

The relationship between green, lean, and Six Sigma strategies can be better understood by examining the significant role of lean manufacturing as an intermediate strategy for connecting green and Six Sigma (Garza-Reyes 2015). In the GLSS strategy, various practices and tools help each other overcome each strategy's individual limitations and increase value when used in different combinations (Gandhi and Thanki 2025; Ghasemibojd et al. 2025; Shokri and Li 2020). Prior studies highlighted various GLSS frameworks for improving the environmental performance of organizations (Kaswan et al. 2024; McGlinchey et al. 2025; Mohan et al. 2025; Rathi et al. 2022; Yadav and Gahlot 2022). In this regard, a study conducted in the construction industry revealed the potential benefits of the GLSS strategy in the form of improved quality and reduced process and environmental waste (Banawi and Bilec 2014). The study concluded by developing a framework combining tools such as LCA, VSM, and cause-and-effect diagrams. Another study emphasized GLSS application in the manufacturing industry and developed a framework comprising the drivers, barriers, and key performance measures of a GLSS strategy (Kumar et al. 2015). In addition, Erdil et al. (2018) developed a framework incorporating the sustainability objectives in the GLSS projects.

Another study presented a conceptual model for small and medium-sized enterprises (SMEs) in India and included the GLSS tools, challenges, critical success factors, and advantages of the GLSS strategy (Siegel et al. 2019). Prior studies also utilized the DMAIC methodology for developing an integrated GLSS framework (Jamil et al. 2020; Kaswan and Rathi 2020; Kaswan et al. 2021; Prashar 2020;

Tsarouhas and Sidiropoulou 2024). For example, a study presented a DMAIC-based GLSS framework based on the extensive review of literature and incorporated various practices and tools at each stage of the DMAIC cycle (Farrukh et al. 2021). The study highlighted the key tools such as supplier-input-process-output-control (SIPOC), VSM, LCA, process mapping, quality function deployment (QFD), and critical to quality (CTQ) at the define stage; LCA, VSM, process mapping, and check sheets at the measure stage, and failure mode and effect analysis (FMEA), check sheets, ANOVA, LCA, 5 whys, brainstorming, Pareto, and cause-effect analysis at the analyse stage. Different tools such as 5S (seiri, seiton, seiso, seiketsu, and shitsuke), LCA, reducing, reusing, and recycling (3R), design for the environment (DFE), Poka-yoke, and cost–benefit analysis are identified as the key ones at the improve stage, followed by the statistical process control (SPC), LCA, VSM, and standard operating procedures (SOP) at the control stage of the DMAIC approach. In another review study, Kaswan et al. (2021) developed a GLSS framework, including the integrated strategies' failure factors and readiness aspects. One study developed a framework based on three dimensions: (i) environmental impact, (ii) economic benefits, and (iii) resource circularity (Lieder and Rashid 2016) to facilitate manufacturing organizations in reducing environmental waste and saving resources. In addition, the relationship between the green, lean, and Six Sigma strategies can also be understood by understanding the similarities among the three strategies. As such, the similarities include waste reduction of energy, water, raw materials, and defective parts (Cherrafi et al. 2017; Gholami et al. 2020; Hussain et al. 2022; Kumar et al. 2016). From the customer perspective, all three strategies share commonalities such as removing non-value-added activities, meeting customers' requirements for eco-friendly products, and achieving customer satisfaction through reducing defective products (Kaswan et al. 2021; Walter et al. 2021).

1.6 Summary

This chapter presents the overview of environmental sustainability issues of manufacturing firms and discusses the sustainable performance. Following this, foundation of GLSS strategy was described including a discussion on individual strategies of green, lean, ad six sigma. In addition, different combinations of these strategies were illustrated such as Green-Lean and Lean -Six Sigma. Then, an integrated approach of the GLSS strategy was explained.

References

Aadithya, B. G., Asokan, P., & Vinodh, S. (2023). Lean manufacturing in fabrication industry: literature review and framework proposal. *International Journal of Quality & Reliability Management*, *40*(6), 1485–1517.

Adeodu, A., Kanakana-Katumba, M. G., & Rendani, M. (2021). Implementation of Lean Six Sigma for production process optimization in a paper production company. *Journal of Industrial Engineering and Management, 14*(3), 661–680.

Ahmed, A., & Mathrani, S. (2019). A Conceptual Framework to Integrate ISO 14001 and Lean for Evaluating Environmental Performance in Meat Industry.

Ahmed, M. (2018). Application of Lean Principles in Five-Star Hotels in Egypt. *International Journal of Heritage, Tourism and Hospitality, 12*(3 (Special Issue)), 181–194.

Al-Rifai, M. H. (2024). Redesigning and optimizing an electronic device assembly cell through lean manufacturing tools and kaizen philosophy: an application case study. *International Journal of Productivity and Performance Management, 73*(4), 1273–1301.

Al-Senani, G. M., Al-Qahtani, S. D., Katamesh, N. S., Ahmed, H. A., & Mohamed, M. A. (2025). Sustainability analysis of RP-UPLC for detecting antidiabetic traces in wastewater: a focus on six sigma. *Green Chemistry Letters and Reviews, 18*(1), 2549735.

Amjad, M. S., Rafique, M. Z., & Khan, M. A. (2021). Leveraging optimized and cleaner production through industry 4.0. *Sustainable Production and Consumption, 26*, 859–871.

Antony, J. (2006). Six sigma for service processes. *Business process Management journal, 12*(2), 234–248.

Antony, J., Rodgers, B., & Cudney, E. A. (2019a). Lean Six Sigma in policing services: case examples, lessons learnt and directions for future research. *Total Quality Management & Business Excellence, 30*(5–6), 613–625.

Antony, J., Sony, M., Dempsey, M., Brennan, A., Farrington, T., & Cudney, E. A. (2019b). An evaluation into the limitations and emerging trends of Six Sigma: an empirical study. *The TQM Journal, 31*(2), 205–221. https://doi.org/10.1108/tqm-12-2018-0191

Appiah, L. O., Essuman, D., Forson, C. A., Boso, N., & Annan, J. (2025). Green process innovation and financial performance in small and medium-sized enterprises in a developing Country: Role of resource orchestration. *Journal of Business Research, 189*, 115210.

Austin, W., & Adebayo, F. (2021). Lean Implementation Barriers and the duality of lean in an Organisations. *Applied Journal of Economics, Management and Social Sciences, 2*(1), 1–8.

Banawi, A., & Bilec, M. M. (2014). A framework to improve construction processes: Integrating Lean, Green and Six Sigma. *International Journal of Construction Management, 14*(1), 45–55. https://doi.org/10.1080/15623599.2013.875266

Bapat, G. S., Kumar, A., Kumar, A., Hota, S. L., Kavita, & Singh, K. (2024). Sustainable Green Manufacturing Approaches in India—A Step towards a New Green Revolution through SMEs. *Engineering Proceedings, 59*(1), 211.

Belamkar, N. (2016). development of green high performance liquid chromatographymethod for determination for ranitidine hydrochloride in solid dosage form by applying lean sigma principles. *Research Journal of Pharmaceutical, Biological and Chemical Sciences.*

Bhat, S., Gijo, E. V., Rego, A. M., & Bhat, V. S. (2020). Lean Six Sigma competitiveness for micro, small and medium enterprises (MSME): an action research in the Indian context. *The TQM Journal, 33*(2), 379–406. https://doi.org/10.1108/tqm-04-2020-0079

Brun, A., & Cicculo, F. (2022). Factors affecting sustainability-oriented innovation in the leather supply chain. *Strategic Change, 31*(3), 305–321.

Burghall, R., Grant, V., & Morgan, J. (2014). *Lean six sigma business transformation for dummies.* John Wiley & Sons.

Cao, Y., Lawson, B., & Pil, F. K. (2023). Social sustainability and human rights in global supply chains. *International Journal of Operations & Production Management, 44*(1), 370–390. https://doi.org/10.1108/ijopm-10-2022-0670

Carrijo, P. R. S., Rader, M. L. B., Batalha, M. O., & Godinho Filho, M. (2024). Lean manufacturing in agriculture: adapting the value stream mapping approach for farm management. *Operations Management Research*, 1–25.

Carvalho, H., Duarte, S., & Cruz Machado, V. (2011). Lean, agile, resilient and green: divergencies and synergies. *International Journal of Lean Six Sigma, 2*(2), 151–179.

Chaudhary, R., Jhanwar, P., & Mudimela, P. R. (2023). Finite element analysis of graphene oxide hinge structure-based RF NEM switch. *IETE Journal of Research*, *69*(2), 967–974.

Chaudhary, R., & Mudimela, P. R. (2022). Comprehensive study of RF analysis of G/GO-based NEMS shunt switch. *Microsystem Technologies*, *28*(4), 1069–1075.

Chen, X., Xu, L., Ren, Z., Jia, F., & Yu, Y. (2023). Sustainable supply chain management in the leather industry: a systematic literature review. *International Journal of Logistics Research and Applications*, *26*(12), 1663–1703.

Cherrafi, A., Elfezazi, S., Chiarini, A., Mokhlis, A., & Benhida, K. (2016). The integration of lean manufacturing, Six Sigma and sustainability: A literature review and future research directions for developing a specific model. *Journal of Cleaner Production*, *139*, 828–846. https://doi.org/10.1016/j.jclepro.2016.08.101

Cherrafi, A., Elfezazi, S., Garza-Reyes, J. A., Benhida, K., & Mokhlis, A. (2017). Barriers in Green Lean implementation: a combined systematic literature review and interpretive structural modelling approach. *Production Planning & Control*, *28*(10), 829–842. https://doi.org/10.1080/09537287.2017.1324184

Chiou, T.-Y., Chan, H. K., Lettice, F., & Chung, S. H. (2011). The influence of greening the suppliers and green innovation on environmental performance and competitive advantage in Taiwan. *Transportation Research Part E: Logistics and Transportation Review*, *47*(6), 822–836.

Chotia, V., Soni, S., Jain, G., & Papa, A. (2023). Barriers for adoption of green supply chain management in cement industry: an interpretive structural modelling (ISM) approach. *Annals of Operations Research*, 1–34.

Chugani, N., Kumar, V., Garza-Reyes, J. A., Rocha-Lona, L., & Upadhyay, A. (2017). Investigating the green impact of Lean, Six Sigma and Lean Six Sigma. *International Journal of Lean Six Sigma*, *8*(1), 7–32. https://doi.org/10.1108/ijlss-11-2015-0043

Ciannella, S., & Santos, L. C. (2022). Exploring the influence of lean manufacturing practices on employee social sustainability. *Social Responsibility Journal*, *18*(8), 1677–1691.

Condé, G. C. P., Oprime, P. C., Pimenta, M. L., Sordan, J. E., & Bueno, C. R. (2023). Defect reduction using DMAIC and Lean Six Sigma: a case study in a manufacturing car parts supplier. *International Journal of Quality & Reliability Management*, *40*(9), 2184–2204.

Crawford, R. (2004), "Ammunition enterprise excellence ready for tomorrow?". *USA Armor School Research Library*, available at: https://www.dtic.mil/ndia/2004munitions/Crawford.pdf/

Cuer, L., Scalco, A. R., & Satolo, E. G. (2025). Lean production in family farming: application of its principles in the production of vegetables. *International Journal of Lean Six Sigma*, *16*(3), 608–634.

Dahiya, A., Kadiyan, S., & Giri, R. (2022). Kaizen and 5S: productivity improvement tools for automotive assembly line. *NOLEGEIN Journal of Operations Research & Management*, *5*(1), 1–10p.

Danta, S., & Rath, B. N. (2024). Do Green Patent and Renewable Energy Consumption Matter for Sustainable Green Growth in the African Region? *Economic Papers: A journal of applied economics and policy*, *43*(4), 370–388. https://doi.org/10.1111/1759-3441.12426

de Souza Costa, I., Cardoso de Oliveira Neto, G., & Rodrigues Leite, R. (2021). How does the use of PPC tools/activities improve eco-efficiency? A systematic literature review. *Production Planning & Control*, *32*(7), 526–548.

Deif, A. M. (2011). A system model for green manufacturing. *Journal of Cleaner Production*, *19*(14), 1553–1559.

Drohomeretski, E., Da Costa, S. G., & De Lima, E. P. (2014a). Green supply chain management: Drivers, barriers and practices within the Brazilian automotive industry. *Journal of Manufacturing Technology Management*, *25*(8), 1105–1134.

Drohomeretski, E., Gouvea da Costa, S. E., Pinheiro de Lima, E., & Garbuio, P. A. D. R. (2014b). Lean, Six Sigma and Lean Six Sigma: an analysis based on operations strategy. *International Journal of Production Research*, *52*(3), 804–824.

Duarte, S., & Cruz-Machado, V. (2013). Modelling lean and green: a review from business models. *International Journal of Lean Six Sigma*, *4*(3), 228–250.

Elkington, J. (1998). Partnerships from cannibals with forks: The triple bottom line of 21st-century business. *Environmental Quality Management*, *8*(1), 37–51.
EPA. (2007). The lean and environmental toolkit. In.
EPA, U. (2003). Lean manufacturing and the environment: research on advanced manufacturing systems and the environment and recommendations for leveraging better environmental performance. *United States Environmental Protection Agency.*
Erdil, N. O., Aktas, C. B., & Arani, O. M. (2018). Embedding sustainability in lean six sigma efforts. *Journal of Cleaner Production, 198*, 520–529. https://doi.org/10.1016/j.jclepro.2018.07.048
Esty, D. C., & Winston, A. (2009). *Green to gold: How smart companies use environmental strategy to innovate, create value, and build competitive advantage.* John Wiley & Sons.
Farrukh, A. (2025). A comparative exploratory study of the critical success factors of a green-lean-six sigma strategy. *Business process Management journal*, 1–35.
Farrukh, A., Mathrani, S., & Sajjad, A. (2021). A DMAIC approach to investigate the green lean six sigma tools for improving environmental performance. 2021 IEEE Asia-Pacific Conference on Computer Science and Data Engineering (CSDE).
Farrukh, A., Mathrani, S., & Sajjad, A. (2023). Green-lean-six sigma practices and supporting factors for transitioning towards circular economy: A natural resource and intellectual capital-based view. *Resources Policy, 84.* https://doi.org/10.1016/j.resourpol.2023.103789
Farrukh, A., Mathrani, S., & Taskin, N. (2020). Investigating the Theoretical Constructs of a Green Lean Six Sigma Approach towards Environmental Sustainability: A Systematic Literature Review and Future Directions. *Sustainability, 12*(19). https://doi.org/10.3390/su12198247
Farrukh, A., Mehddi, A., Sajjad, A., & Yawar, S. A. (2025). Barriers to women's participation and contribution in the process manufacturing industry: Managerial insights from multiple case studies in an emerging economy. *Journal of Management & Organization*, 1–31. https://doi.org/10.1017/jmo.2025.10045
Faulkner, W., & Badurdeen, F. (2014). Sustainable Value Stream Mapping (Sus-VSM): methodology to visualize and assess manufacturing sustainability performance. *Journal of Cleaner Production, 85*, 8–18. https://doi.org/10.1016/j.jclepro.2014.05.042
Faulks, B., Song, Y., Waiganjo, M., Obrenovic, B., & Godinic, D. (2021). Impact of empowering leadership, innovative work, and organizational learning readiness on sustainable economic performance: an empirical study of companies in Russia during the COVID-19 pandemic. *Sustainability, 13*(22), 12465.
Fayyaz, A., Liu, C., Xu, Y., Farooq, A., & Ahmed, S. (2025). Linkages between lean six sigma, green manufacturing, circular economy and operational performance in manufacturing organizations of a developing country. *International Journal of Lean Six Sigma.*
Fercoq, A., Lamouri, S., & Carbone, V. (2016). Lean/Green integration focused on waste reduction techniques. *Journal of Cleaner Production, 137*, 567–578.
Gaikwad, L., & Sunnapwar, V. (2020). Development of an integrated framework of LGSS strategies for Indian manufacturing firms to improve business performance: an empirical study. *The TQM Journal, 33*(1), 257–291. https://doi.org/10.1108/tqm-05-2020-0110
Gandhi, J. D., & Thanki, S. (2025). Sustainability index development by integrating lean green and Six Sigma tools: a case study of the Indian manufacturing industry. *International Journal of Productivity and Performance Management, 74*(3), 793–818.
Gao, S., Low, S. P., & Tok, S. K. W. (2025). Applications of lean principles in the operations of architectural and engineering design firms: an exploratory study in Singapore. *Engineering, Construction and Architectural Management.*
Garza-Reyes, J. A. (2015). Green lean and the need for Six Sigma. *International Journal of Lean Six Sigma, 6*(3), 226–248. https://doi.org/10.1108/ijlss-04-2014-0010
Gebeyehu, S. G., Abebe, M., & Gochel, A. (2022). Production lead time improvement through lean manufacturing. *Cogent Engineering, 9*(1), 2034255.
Ghasemibojd, F., Franchetti, M. J., & George, B. (2025). Green lean six sigma for sustainable development: a systematic review of evolution, challenges, and future pathways. *Clean Technologies and Environmental Policy.* https://doi.org/10.1007/s10098-025-03281-y

Gholami, H., Jamil, N., Mat Saman, M. Z., Streimikiene, D., Sharif, S., & Zakuan, N. (2020). The application of Green Lean Six Sigma. *Business Strategy and the Environment.*

Ghosh, M. (2013). Lean manufacturing performance in Indian manufacturing plants. *Journal of Manufacturing Technology Management*, *24*(1), 113–122.

Glavič, P., & Lukman, R. (2007). Review of sustainability terms and their definitions. *Journal of Cleaner Production*, 15(18), 1875–1885.

Gochel, A., Gebeyehu, S. G., & Abebe, M. (2022). Production lead time improvement through lean manufacturing. *International Journal of Industrial and Systems Engineering*, *40*(2), 147–156.

Hariyani, D., & Mishra, S. (2022a). Barriers to the adoption of integrated sustainable-green-lean-six sigma-agile manufacturing system (ISGLSAMS): a literature review. *Benchmarking: An International Journal*, *30*(9), 3590–3636. https://doi.org/10.1108/bij-10-2021-0585

Hariyani, D., & Mishra, S. (2022b). Drivers for the adoption of integrated sustainable green lean six sigma agile manufacturing system (ISGLSAMS) and research directions. *Cleaner engineering and technology*, *7*, 100449.

Hariyani, D., Mishra, S., Sharma, M. K., & Hariyani, P. (2023). A study of the barriers to the adoption of integrated sustainable-green-lean-six sigma-agile manufacturing system (ISGLSAMS) in Indian manufacturing organizations. *Cleaner Waste Systems*, *5*. https://doi.org/10.1016/j.clwas.2023.100098

Hasimoglu, S. (2021). Impact of Innovated Per Adult Human Unit Method (PAHUM) vs. Error Bound Per Capita (PC): Recalibration of EU and Developing Countries (India), Family/Households' Population, Food Security and Environmental Sustainability is ACCEPTED for publication as a book chapter in the following book: Challenging Issues on Environment and Earth Science Vol. 5.(International Book). F. No. *Challenging Issues on Environment and Earth Science Vol. 5*, 78–96.

Helmold, M., Küçük Yılmaz, A., Flouris, T., Winner, T., Cvetkoska, V., & Dathe, T. (2022). Lean Management: Introduction. In *Lean Management, Kaizen, Kata and Keiretsu: Best-Practice Examples and Industry Insights from Japanese Concepts* (pp. 1–17). Springer.

Hilton, R. J., & Sohal, A. (2012). A conceptual model for the successful deployment of Lean Six Sigma. *International Journal of Quality & Reliability Management*, *29*(1), 54–70.

Hofer, C., Eroglu, C., & Hofer, A. R. (2012). The effect of lean production on financial performance: The mediating role of inventory leanness. *International Journal of Production Economics*, *138*(2), 242–253.

Holweg, M. (2007). The genealogy of lean production. *Journal of Operations Management*, *25*(2), 420–437.

Hoogmartens, R., Van Passel, S., Van Acker, K., & Dubois, M. (2014). Bridging the gap between LCA, LCC and CBA as sustainability assessment tools. *Environmental Impact Assessment Review*, *48*, 27–33. https://doi.org/10.1016/j.eiar.2014.05.001

Huo, B., Gu, M., & Wang, Z. (2019). Green or lean? A supply chain approach to sustainable performance. *Journal of Cleaner Production*, *216*, 152–166.

Hussain, K., He, Z., Ahmad, N., Iqbal, M., & Saeed, M. Z. (2022). Establishing a Green, Lean and Six Sigma implementation model for sustainable construction industry: an analysis of driving forces through ISM-MICMAC approach. *Environ Sci Pollut Res Int*. https://doi.org/10.1007/s11356-022-24039-9

Ijaz, T., Qazalbash, A. H. R., Razzaq, A. A., Rafique, M. Z., Khan, M. A., & Jamshaid, S. H. (2024). Development of green manufacturing implementation framework based on life-cycle assessment. *Sustainable Environment Research*, *34*(1), 23.

Ishak, A., Mohamad, E., & Rahman, M. A. A. (2024). Confirmatory factor analysis for the implementation of Green Lean Six Sigma (GLSS) in Malaysia's wastewater treatment industry. *Water Policy*. https://doi.org/10.2166/wp.2024.095

Jamil, N., Gholami, H., Saman, M. Z. M., Streimikiene, D., Sharif, S., & Zakuan, N. (2020). DMAIC-based approach to sustainable value stream mapping: towards a sustainable manufacturing system. *Economic Research-Ekonomska Istraživanja*, *33*(1), 331–360. https://doi.org/10.1080/1331677x.2020.1715236

Jamil, N., Hasan, M. Z., Yaacob, T. Z., Omain, S. Z., & Zakuan, N. Integrating Green Lean Six Sigma for Sustainable Operational Excellence: A Review of Applications in Malaysia's Food and Beverage Industry.

Jaskari, S. (2014). Applying Lean-Principles for Electric Motors Production.

Javed, M. O. A. (2025). Implementing Green Lean Six Sigma: a methodology for promoting sustainability. *International Journal of Productivity and Quality Management*, *45*(3), 334–354.

Jayanth, B. V., Prathap, P., Sivaraman, P., Yogesh, S., & Madhu, S. (2020). Implementation of lean manufacturing in electronics industry. *Materials Today: Proceedings*, *33*, 23–28.

Jin, H., Qu, P., Zhou, Y., Qu, Q., & Li, H. (2024). Green Lean Six Sigma practice in the industrial and service sectors: from the systematic literature review to an integration diagram. *Production Planning & Control*, 1–26. https://doi.org/10.1080/09537287.2024.2369779

Joseph, R. E., Kanya, N., Bhaskar, K., Xavier, J. F., Sendilvelan, S., Prabhahar, M.,...Geetha, S. (2021). Analysis on productivity improvement, using lean manufacturing concept. *Materials Today: Proceedings*, *45*, 7176–7182.

Julie, S., Potter, A., & Geng, R. (2024). Examining the effects of stakeholder forces on sustainable practices in the Bangladeshi garment industry. *Cleaner Logistics and Supply Chain*, *12*. https://doi.org/10.1016/j.clscn.2024.100162

Karim, A., & Arif-Uz-Zaman, K. (2013). A methodology for effective implementation of lean strategies and its performance evaluation in manufacturing organizations. *Business process Management journal*, *19*(1), 169–196.

Kaswan, M. S., & Rathi, R. (2020). Green Lean Six Sigma for sustainable development: Integration and framework. *Environmental Impact Assessment Review*, *83*. https://doi.org/10.1016/j.eiar.2020.106396

Kaswan, M. S., Rathi, R., Antony, J., Cross, J., Garza-Reyes, J. A., Singh, M.,...Sony, M. (2024). Integrated Green Lean Six Sigma-Industry 4.0 approach to combat COVID-19: from literature review to framework development. *International Journal of Lean Six Sigma*, *15*(1), 50–79. https://doi.org/10.1108/IJLSS-11-2022-0227

Kaswan, M. S., Rathi, R., Reyes, J. A. G., & Antony, J. (2021). Exploration and Investigation of Green Lean Six Sigma Adoption Barriers for Manufacturing Sustainability. *IEEE Transactions on Engineering Management*, 1–15. https://doi.org/10.1109/tem.2021.3108171

Khan, F., & Ali, Y. (2022). Implementation of the circular supply chain management in the pharmaceutical industry. *Environ Dev Sustain*, *24*(12), 13705–13731. https://doi.org/10.1007/s10668-021-02007-6

Khan, S. A. R., Yu, Z., & Farooq, K. (2023). Green capabilities, green purchasing, and triple bottom line performance: Leading toward environmental sustainability. *Business Strategy and the Environment*, *32*(4), 2022–2034.

Kharat, M. G., Kapoor, S., Parhi, S., Kharat, M. G., & Pandey, S. (2025). Operationalizing sustainability in pharmaceuticals: Green supply chain metrics for circular economy. *Sustainable Futures*, *9*. https://doi.org/10.1016/j.sftr.2024.100413

Klein, L. L., Moyano-Fuentes, J., Vieira, K. M., & Marçal, D. R. (2024). An exploratory study of the relationships between Lean practices and team performance in higher education. *International Journal of Lean Six Sigma*, *15*(2), 372–399. https://doi.org/10.1108/IJLSS-08-2022-0168

Kumar, S., Kumar, N., & Haleem, A. (2015). Conceptualisation of sustainable green lean six sigma: an empirical analysis. *International Journal of Business Excellence*, *8*(2), 210–250.

Kumar, S., Luthra, S., Govindan, K., Kumar, N., & Haleem, A. (2016). Barriers in green lean six sigma product development process: an ISM approach. *Production Planning & Control*, 1–17. https://doi.org/10.1080/09537287.2016.1165307

Kumar, S., Satsangi, P. S., & Prajapati, D. R. (2013). Improvement of Sigma level of a foundry: a case study. *The TQM Journal*, *25*(1), 29–43. https://doi.org/10.1108/17542731311286414

Kwon, H.-J. (2020). Climate change and health: more research is still needed. *Journal of Preventive Medicine and Public Health*, *53*(1), 1.

Lieder, M., & Rashid, A. (2016). Towards circular economy implementation: a comprehensive review in context of manufacturing industry. *Journal of Cleaner Production, 115*, 36–51. https://doi.org/10.1016/j.jclepro.2015.12.042

Liew, W. T., Adhitya, A., & Srinivasan, R. (2014). Sustainability trends in the process industries: A text mining-based analysis. *Computers in Industry, 65*(3), 393–400. https://doi.org/10.1016/j.compind.2014.01.004

Lizarelli, F. L., Chakraborty, A., Antony, J., Jayaraman, R., Carneiro, M. B., & Furterer, S. (2023). Lean and its impact on sustainability performance in service companies: results from a pilot study. *The TQM Journal, 35*(3), 698–718.

Lu, C., & Wang, K. (2023). Natural resource conservation outpaces and climate change: Roles of reforestation, mineral extraction, and natural resources depletion. *Resources Policy, 86*. https://doi.org/10.1016/j.resourpol.2023.104159

Lu, L.-H., Wang, Z., Xia, W., Cheng, P., Zhang, B., Cao, Z., & He, W.-M. (2019). Sustainable routes for quantitative green selenocyanation of activated alkynes. *Chinese Chemical Letters, 30*(6), 1237–1240.

Lu, S., Hou, W., Li, Z., Xia, J., Xie, L., & Su, H. (2022). Design and optimisation of a green manufacturing-recycling network considering heavy metal pollutants–an electronic assembly case. *International Journal of Production Research, 60*(9), 2830–2849.

Luoma, P., Penttinen, E., Tapio, P., & Toppinen, A. (2022). Future images of data in circular economy for textiles. *Technological Forecasting and Social Change, 182*, 121859. https://doi.org/10.1016/j.techfore.2022.121859

Mahmoud, M. A., Tsetse, E. K. K., Tulasi, E. E., & Muddey, D. K. (2022). Green packaging, environmental awareness, willingness to pay and consumers' purchase decisions. *Sustainability, 14*(23), 16091.

Maleyeff, J., Arnheiter, E. A., & Venkateswaran, V. (2012). The continuing evolution of Lean Six Sigma. *The TQM Journal, 24*(6), 542–555. https://doi.org/10.1108/17542731211270106

Marco-Ferreira, A., Stefanelli, N. O., Seles, B. M. R. P., & Fidelis, R. (2020). Lean and Green: practices, paradigms and future prospects. *Benchmarking: An International Journal, 27*(7), 2077–2107.

Marrucci, L., Marchi, M., & Daddi, T. (2020). Improving the carbon footprint of food and packaging waste management in a supermarket of the Italian retail sector. *Waste Manag, 105*, 594–603. https://doi.org/10.1016/j.wasman.2020.03.002

Martins, A. d. O., dos Anjos, F. E. V., & da Silva, D. O. (2023). The lean farm: application of tools and concepts of lean manufacturing in agro-pastoral crops. *Sustainability, 15*(3), 2597.

Mathiyazhagan, K., Agarwal, V., Appolloni, A., Saikouk, T., & Gnanavelbabu, A. (2021). Integrating lean and agile practices for achieving global sustainability goals in Indian manufacturing industries. *Technological Forecasting and Social Change, 171*, 120982.

McGlinchey, Y., Iqbal, J., Trubetskaya, A., Thenarasu, M., & McDermott, O. (2025). Design for green lean six sigma to improve sustainability in the pharmaceutical industry—A case study. *Production & Manufacturing Research, 13*(1). https://doi.org/10.1080/21693277.2025.2461010

Middleton, P., & Sutton, J. (2020). Lean Production—Five Principles. In *Lean Software Strategies* (pp. 21–48). Productivity Press.

Mishra, R., Singh, R. K., & Garza-Reyes, J. A. (2023). Interplay between absorptive capacity, analytics competence and sustainable economic performance of MSMEs in supply chain: The mediating role of risk resilience. *Annals of Operations Research*, 1–25.

Mohamad, N., Muthusamy, K., Embong, R., Kusbiantoro, A., & Hashim, M. H. (2022). Environmental impact of cement production and Solutions: A review. *Materials Today: Proceedings*, 48, 741–746.

Mohan, J., Kaswan, M. S., & Rathi, R. (2024). An analysis of green lean six sigma deployment in MSMEs: a systematic literature review and conceptual implementation framework. *The TQM Journal*.

Mohan, J., Kaswan, M. S., & Rathi, R. (2025). An analysis of green lean six sigma deployment in MSMEs: a systematic literature review and conceptual implementation framework. *The TQM Journal*, *37*(3), 747–777. https://doi.org/10.1108/TQM-06-2023-0197

Mohan, S. V., Katakojwala, R. J. C. O. i. G., & Chemistry, S. (2021). The circular chemistry conceptual framework: A way forward to sustainability in industry 4.0. *28*, 100434.

Mu, Y., Duan, C., Li, X., & Wu, Y. (2023). A Monitoring Method for Corporate Environmental Performance Based on Data Fusion in China under the Double Carbon Target. *Sustainability*, *15*(12), 9391.

Nadeem, S. P., Garza-Reyes, J. A., Anosike, A. I., & Kumar, V. (2019). Coalescing the lean and circular economy.

Narassima, M. S., Aashrith, V., Aldo Ronald, C., Anbuudayasankar, S. P., & Thenarasu, M. (2025). Implementation of lean principles in the yarn manufacturing industry: a system dynamics approach. Benchmarking: An International Journal, 32(1), 222–240.

Neubaum, D. O., & Zahra, S. A. (2006). Institutional ownership and corporate social performance: The moderating effects of investment horizon, activism, and coordination. *Journal of Management*, 32(1), 108–131.

Ng, R., Low, J. S. C., & Song, B. (2015). Integrating and implementing Lean and Green practices based on proposition of Carbon-Value Efficiency metric. *Journal of Cleaner Production, 95*, 242–255. https://doi.org/10.1016/j.jclepro.2015.02.043

Nishad, S., Sahu, A. K., & Sahu, N. K. (2024). Advocating Lean Practices and Strategies in Decision-Making for Reinforcing Industrial and Manufacturing Designs. *Industrial and Manufacturing Designs: Quantitative and Qualitative Analysis*, 75–103.

Nutsugah, F. F., Anning-Dorson, T., Braimah, S. M., & Tweneboah-Koduah, E. Y. (2021). Candle under a bushel: communicating environmental performance to improve firm performance. *International Journal of Productivity and Performance Management*, *70*(8), 1953–1971.

Omoloso, O., Mortimer, K., Wise, W. R., & Jraisat, L. (2021). Sustainability research in the leather industry: A critical review of progress and opportunities for future research. *Journal of Cleaner Production, 285*, 125441.

Owusu-Manu, D.-G., Babon-Ayeng, P., Kissi, E., Edwards, D. J., Okyere-Antwi, D., & Elgohary, H. (2023). Green construction and environmental performance: an assessment framework. *Smart and Sustainable Built Environment*, *12*(3), 565–583.

Pampanelli, A. B., Found, P., & Bernardes, A. M. (2014). A Lean & Green Model for a production cell. *Journal of Cleaner Production*, *85*, 19–30. https://doi.org/10.1016/j.jclepro.2013.06.014

Panagiotopoulou, V. C., Stavropoulos, P., & Chryssolouris, G. (2022). A critical review on the environmental impact of manufacturing: a holistic perspective. *The International Journal of Advanced Manufacturing Technology*, 118(1), 603–625.

Pande, P., Neuman, R., & Cavanagh, R. (2000). *The Six Sigma Way, Chapter 12-Identifying Core Processes and Key Customers.* McGraw Hill Professional.

Pandey, H., Garg, D., & Luthra, S. (2018). Identification and ranking of enablers of green lean Six Sigma implementation using AHP. *International Journal of Productivity and Quality Management*, *23*(2), 187–217.

Papafloratos, T., Markidis, I., Kotzaivazoglou, I., & Fragidis, G. J. S. (2023). Sustainability material topics and materiality analysis in the chemical industry. *15*(18), 14014.

Patel, M., & Desai, D. A. (2018). Critical review and analysis of measuring the success of Six Sigma implementation in manufacturing sector. *International Journal of Quality & Reliability Management*, *35*(8), 1519–1545. https://doi.org/10.1108/ijqrm-04-2017-0081

Pepper, M. P. J., & Spedding, T. A. (2010). The evolution of lean Six Sigma. *International Journal of Quality & Reliability Management*, *27*(2), 138–155.

Pingyu, Y., & Yu, Y. (2010). The barriers to SMEs' implementation of lean production and countermeasures: based on SMS in Wenzhou. *International Journal of Innovation, Management and Technology*, *1*(2), 220–225.

Pojasek, R. B. (2003). Lean, six sigma, and the systems approach: Management initiatives for process improvement. *Environmental Quality Management*, *13*(2), 85–92.

Prashar, A. (2020). Adopting Six Sigma DMAIC for environmental considerations in process industry environment. *The TQM Journal*, *ahead-of-print*(ahead-of-print). https://doi.org/10.1108/tqm-09-2019-0226

Raduan, M. I. M., Muhammad, N. A., Ab-Samat, H., Abdullah, J., Jeng, F. C., Amar, K., & Muhammad, N. (2025). The Impact of Lean Six Sigma on Organizational Sustainability. *Semarak Advanced Research in Organizational Behaviour*, *5*(1), 17–33.

Rahman, H. U., Zahid, M., Ullah, M., & Al-Faryan, M. A. S. (2023). Green supply chain management and firm sustainable performance: The awareness of China Pakistan Economic Corridor. *Journal of Cleaner Production*, *414*, 137502.

Rasheed, R., Rashid, A., & Ngah, A. H. (2025). Role of leadership styles to foster innovative capabilities and green purchasing. *Journal of Global Operations and Strategic Sourcing*, *18*(2), 328–348.

Rashid, A., Rasheed, R., & Altay, N. (2025). Greening manufacturing: the role of institutional pressure and collaboration in operational performance. *Journal of Manufacturing Technology Management*, *36*(2), 455–478.

Rathi, R., Kaswan, M. S., Garza-Reyes, J. A., Antony, J., & Cross, J. (2022). Green Lean Six Sigma for improving manufacturing sustainability: Framework development and validation. *Journal of Cleaner Production*, *345*. https://doi.org/10.1016/j.jclepro.2022.131130

Rathi, R., Vakharia, A., & Kaswan, M. S. (2021). Grey relational analysis of Green Lean Six Sigma critical success factors for improved organisational performance. *International Journal of Six Sigma and Competitive Advantage*, *13*(1–3), 55–75.

Robinson, B. H. (2009). E-waste: an assessment of global production and environmental impacts. *Science of the Total Environment*, 408(2), 183–191.

Ruel, S., & Fritz, M. M. C. (2021). Gender diversity in supply chains: towards more sustainable decisions? evidence from interviews. *Supply Chain Forum: An International Journal*, *22*(3), 205–222. https://doi.org/10.1080/16258312.2021.1948307

Rusinko, C. (2007). Green manufacturing: an evaluation of environmentally sustainable manufacturing practices and their impact on competitive outcomes. *IEEE Transactions on Engineering Management*, *54*(3), 445–454.

Sadikin, M. A. (2023). Defect reduction in the manufacturing industry: Systematic literature review. *International Journal of Industrial Engineering and Engineering Management*, *5*(2), 73–83.

Sadiku, M. N. O., Ajayi-Majebi, A. J., & Adebo, P. O. (2023). Green manufacturing. In *Emerging technologies in manufacturing* (pp. 123–140). Springer.

Sagnak, M., & Kazancoglu, Y. (2016). Integration of green lean approach with six sigma: an application for flue gas emissions. *Journal of Cleaner Production*, *127*, 112–118. https://doi.org/10.1016/j.jclepro.2016.04.016

Salah, S., Rahim, A., & Carretero, J. A. (2010). The integration of Six Sigma and lean management. *International Journal of Lean Six Sigma*, *1*(3), 249–274.

Salifu, S., Seidu, A., & Fuseini, M. A.-B. (2025). Key Green Supply Chain Management Practices Today. In *Green Supply Chain Management* (pp. 35–55). Apple Academic Press.

Setiawan, S. A. (2025). Implementation of Six Sigma methodology to reduce high defect rate in rubber processing industry. *European Journal of Business and Management Research*, *10*(1), 118–126.

Shabeen, S. R., & Krishnan, K. A. (2022). Application of lean manufacturing using value stream mapping (VSM) in precast component manufacturing: A case study. *Materials Today: Proceedings*, *65*, 1105–1111.

Shang, K.-C., Lu, C.-S., & Li, S. (2010). A taxonomy of green supply chain management capability among electronics-related manufacturing firms in Taiwan. *Journal of Environmental Management*, *91*(5), 1218–1226.

Sharma, D., & Singh, D. (2025). Green Lean Six Sigma Barriers and Efficacy in Ceramic Manufacturing Industries: A PLS-SEM Study.

Sharma, V. K., Chandna, P., & Bhardwaj, A. (2017). Green supply chain management related performance indicators in agro industry: A review. *Journal of Cleaner Production, 141*, 1194–1208.

Shokri, A., & Li, G. (2020). Green implementation of Lean Six Sigma projects in the manufacturing sector. *International Journal of Lean Six Sigma, 11*(4), 711–729.

Shou, Y., Shao, J., Lai, K.-h., Kang, M., & Park, Y. (2019). The impact of sustainability and operations orientations on sustainable supply management and the triple bottom line. *Journal of Cleaner Production, 240*. https://doi.org/10.1016/j.jclepro.2019.118280

Siefan, S., Antony, J., Mayyas, A., Omar, M., Swarnakar, V., Lameijer, B.,…Maalouf, M. (2025). The impact of operational excellence methodologies on sustainable performance in the service sector: a systematic literature review. *Production Planning & Control*, 1–28. https://doi.org/10.1080/09537287.2024.2443073

Siegel, R., Antony, J., Garza-Reyes, J. A., Cherrafi, A., & Lameijer, B. (2019). Integrated green lean approach and sustainability for SMEs: From literature review to a conceptual framework. *Journal of Cleaner Production, 240*, 118205.

Silva, M. E., Ruel, S., & Sousa-Filho, J. M. (2023). Measuring supplier diversity, equity and inclusion (DEI): scale development and empirical validation. *Supply Chain Management: An International Journal*. https://doi.org/10.1108/scm-06-2023-0306

Simões, J. M. S., Toledo, J. C. d., & Lizarelli, F. L. (2024). Front-line lean leader capacities, practices and effects on implementing tools: a survey of leaders in industrial companies. *International Journal of Lean Six Sigma, 15*(4), 925–956.

Singh, M., Rathi, R., & Garza-Reyes, J. A. (2021). Analysis and prioritization of Lean Six Sigma enablers with environmental facets using best worst method: A case of Indian MSMEs. *Journal of Cleaner Production, 279*, 123592.

Slavina, T., & Štefanić, N. (2024). Facing challenges of implementing total productive management and lean tools in manufacturing enterprises. *Systems, 12*(2), 52.

Snee, R. D. (2004). Six–Sigma: the evolution of 100 years of business improvement methodology. *International Journal of Six Sigma and Competitive Advantage, 1*(1), 4–20.

Sohani, A., Vrchota, J., Türkmen, B. A., Hosseini, M. M., & Panchal, H. (2022). Green manufacturing and environment. In *Materials for Lightweight Constructions* (pp. 111–135). CRC Press.

Sony, M., Antony, J., Park, S., & Mutingi, M. (2019). Key Criticisms of Six Sigma: A Systematic Literature Review. *IEEE Transactions on Engineering Management*, 1–13. https://doi.org/10.1109/tem.2018.2889517

Sony, M., & Naik, S. (2019). Green Lean Six Sigma implementation framework: a case of reducing graphite and dust pollution. *International Journal of Sustainable Engineering*, 1–10. https://doi.org/10.1080/19397038.2019.1695015

Sordan, J. E. E., Oprime, P. C., Ferreira, J. L., Marinho, C. A., & Pata, A. (2025). Lean manufacturing for reducing lead time in foundry processes: a design science approach. *International Journal of Lean Six Sigma, 16*(2), 328–345.

Sreedharan V, R., G, S., & Raju, R. (2018). Development of a Green Lean Six Sigma model for public sectors. *International Journal of Lean Six Sigma, 9*(2), 238–255. https://doi.org/10.1108/ijlss-02-2017-0020

Sultana, R., Dwivedi, A., & Moktadir, M. A. (2023). Investigating the role of consumers, producers, and policymakers: A case of leather supply chain towards sustainable chemistry. *Current Opinion in Green and Sustainable Chemistry, 39*, 100724.

Sun, L.-y., Miao, C.-l., & Yang, L. (2017). Ecological-economic efficiency evaluation of green technology innovation in strategic emerging industries based on entropy weighted TOPSIS method. *Ecological Indicators, 73*, 554–558.

Sunder M, V. (2016). Lean six sigma project management–a stakeholder management perspective. *The TQM Journal, 28*(1), 132–150.

Székely, F., & Knirsch, M. (2005). Responsible leadership and corporate social responsibility: Metrics for sustainable performance. *European Management Journal*, 23(6), 628–647.

Tsarouhas, P., & Sidiropoulou, N. (2024). Application of Six Sigma methodology using DMAIC approach for a packaging olives production system: a case study. *International Journal of Lean Six Sigma, 15*(2), 247–273. https://doi.org/10.1108/IJLSS-06-2022-0140

Tsolakis, N., Bam, W., Srai, J. S., & Kumar, M. (2019). Renewable chemical feedstock supply network design: The case of terpenes. *Journal of Cleaner Production, 222*, 802–822. https://doi.org/10.1016/j.jclepro.2019.02.108

Turker, D., & Altuntas, C. (2014). Sustainable supply chain management in the fast fashion industry: An analysis of corporate reports. *European Management Journal, 32*(5), 837–849. https://doi.org/10.1016/j.emj.2014.02.001

United Nations Development Programme (2020), UNDP annual report 2020. Available from: https://annualreport.undp.org/2020/assets/UNDP-Annual-Report-2020-en.pdf

Walter, O. M. F. C., Paladini, E. P., Henning, E., & Kalbusch, A. (2021). Recent developments in sustainable lean six sigma frameworks: literature review and directions. *Production Planning & Control*, 1–19. https://doi.org/10.1080/09537287.2021.1967497

Wang, C.-H., Chen, K.-S., & Tan, K.-H. (2019). Lean Six Sigma applied to process performance and improvement model for the development of electric scooter water-cooling green motor assembly. *Production Planning & Control, 30*(5–6), 400–412.

Womack, J. P., Jones, D. T., & Roos, D. (2007). *The machine that changed the world: The story of lean production—Toyota's secret weapon in the global car wars that is now revolutionizing world industry*. Simon and Schuster.

Wong, W. P., & Wong, K. Y. (2014). Synergizing an ecosphere of lean for sustainable operations. *Journal of Cleaner Production, 85*, 51–66.

Xing, Y., & Liu, Y. (2023). Integrating product-service innovation into green supply chain management from a life cycle perspective: A systematic review and future research directions. *Technovation, 126*, 102825.

Yadav, V., & Gahlot, P. (2022). Green Lean Six Sigma sustainability-oriented framework for small and medium enterprises. *International Journal of Quality & Reliability Management, 39*(7), 1787–1807.

Yadav, V., Gahlot, P., Kaswan, M. S., Rathi, R., & Singh, M. (2021a). Sustainable green lean six sigma methodology and application status: a perspective review. *Recent Trends in Industrial and Production Engineering: Select Proceedings of ICAST 2020*, 251–266.

Yadav, V., Gahlot, P., Rathi, R., Yadav, G., Kumar, A., & Kaswan, M. S. (2021b). Integral measures and framework for green lean six sigma implementation in manufacturing environment. *International Journal of Sustainable Engineering, 14*(6), 1319–1331. https://doi.org/10.1080/19397038.2021.1970855

Yadav, V., Kaswan, M. S., Gahlot, P., Duhan, R. K., Garza-Reyes, J. A., Rathi, R.,…Yadav, G. (2023). Green Lean Six Sigma for sustainability improvement: a systematic review and future research agenda. *International Journal of Lean Six Sigma, 14*(4), 759–790. https://doi.org/10.1108/ijlss-06-2022-0132

Yadav, V., Kumar, V., Gahlot, P., Mittal, A., Kaswan, M. S., Garza-Reyes, J. A.,…Owad, A. A. (2023). Exploration and mitigation of green lean six sigma barriers: a higher education institution's perspective. *The TQM Journal*. https://doi.org/10.1108/tqm-03-2023-0069

Yu, W., & Jin, X. (2022). Long-term and short-term effects of green strategy on corporate performance: evidence from Chinese listed companies. *Environmental Science and Pollution Research, 29*(16), 24112–24130.

WCED. (1987). *Our common future*. New York, NY: Oxford University Press.

Chapter 2
Green-Lean-Six Sigma: Drivers, Barriers, Practices, Tools, and Critical Success Factors

2.1 Introduction

This chapter focuses on salient issues related to the Green-Lean-Six Sigma (GLSS) approach, which is critical for enhancing a firm's environmental sustainability, including drivers, barriers, tools, and critical success factors (CSFs). The chapter discusses various drivers of a GLSS strategy that can guide practitioners in understanding the significance of embracing this approach to managing production operations. Next, understanding the barriers while executing the GLSS approach can help corporate managers carefully examine the possible challenges hindering the adoption of this strategy. Following this, various GLSS tools are discussed that can assist the consultants and practitioners in recognizing the sustainability-related benefits of this strategy. Along with the GLSS tools, developing an in-depth understanding of CSFs is imperative for practitioners to realize the vital aspects for successfully deploying this approach.

2.2 Green-Lean-Six Sigma for Environmental Sustainability

2.2.1 Drivers of Green-Lean-Six Sigma Adoption

GLSS is an emerging strategy, and its implementation in the manufacturing and service industries is relatively new. Hence, it is critical to identify the drivers behind GLSS adoption as a central operational strategy (Shokri et al. 2022) employed to enhance environmental performance (Farrukh et al. 2021b; Kaswan and Rathi 2019). The extant literature revealed various internal and external drivers behind the GLSS adoption in the organizations (Cherrafi et al. 2016a; Farrukh et al. 2021b; Shokri et al. 2022).

A. Farrukh and A. Sajjad, *A Manager's Guide to Green-Lean-Six Sigma Application*,
https://doi.org/10.1007/978-3-032-21111-8_2

Internal drivers include profit margin protection, cost reduction, financial incentives, and competitive advantage behind the GLSS adoption in the manufacturing sector (Albliwi et al. 2015; Cherrafi et al. 2016b; Farrukh et al. 2021b, 2023a; Subramanian and Abdulrahman 2017; Trianni et al. 2016). In addition, an organization's willingness and desire to execute the GLSS strategy are key internal motivators behind the adoption of GLSS (Kaswan and Rathi 2019). Conversely, external drivers include customer pressure, competitive advantage, regulatory requirements, and supplier pressure. As such, the customer pressure towards environmental efforts of manufacturing organizations is an instrumental factor in adopting green practices (Cherrafi et al. 2016a; Famiyeh et al. 2018; Fargani et al. 2016; Mittal and Sangwan 2015; Mollenkopf et al. 2010). Similarly, increasing legislative requirements, including both the current and future regulations, compel firms to adopt eco-friendly practices (Farrukh et al. 2023a; Gandhi et al. 2018; Sajjad et al. 2019; Shokri et al. 2022). Rising awareness concerning environmental issues has also made consumers, non-governmental organizations (NGOs), and the public more vigilant about the adverse impacts of firms' environmental activities and performance (Fargani et al. 2016; Farrukh et al. 2021b; Liu et al. 2021). Additionally, competitive pressures propel manufacturers to adopt environmental management strategies as leading competitors set industry benchmarks and pioneer pro-environmental initiatives that prompt lagging firms to adopt proactive environmental practices (Walker et al. 2008).

From this standpoint, one study identified a list of internal and external drivers in the manufacturing sector in a developed and developing economy context (Farrukh et al. 2021b). The authors conducted a qualitative study using semi-structured interviews with the Lean Six Sigma (LSS) and environmental consultants. Building on the experts' insights, the study highlighted internal drivers, including organizational and operational motivators, where the organizational motivators included continuous improvement, environmental awareness, and resource circulation aspects, and operational drivers comprised factors such as process control, cost reduction, and waste reduction. The findings highlighted environmental cost reduction as the key operational motivator for manufacturers in a developed economy and operational cost reduction as a significant factor in a developing economy. Similarly, resource circulation was identified as a significant organizational driver for initiating the GLSS approach in developed economies.

On the other hand, regulatory, society-related, and market-oriented drivers were synthesized under the external driver category. Among these drivers, environmental regulations in developed and developing countries, including national regulatory requirements, prompting manufacturing organizations to initiate GLSS practices, were considered. Similarly, customer influence, competitor pressure, and company image were collated under the market-oriented drivers, and public pressure and influence of development organizations were combined under the society-oriented drivers. The findings highlighted the stringent environmental legislation as a key regulatory driver in developed economies compared to developing countries. Conversely, competitive pressure was identified as a key market-oriented driver in developing economies, and customer pressure as a significant reason for implementing

GLSS in developed countries. Finally, the international development organizations were recognized as an emerging driver, compelling manufacturers in developing economies to adopt GLSS practices.

Similarly, another study conducted in Morocco investigated the GLSS drivers in diverse industrial settings (Cherrafi et al. 2016b). The authors first developed a framework, building on the extensive literature review. Subsequently, using a longitudinal approach, they implemented the proposed framework in the textile, tannery, hotel, and agri-food sectors. The study revealed key internal drivers: employee satisfaction, firm image, profitability, cost reduction, and process improvement. Similarly, major external drivers comprised stakeholder requirements, consumer demands, regulatory requirements, and competitive advantage. Manufacturing firms are required to follow environmental guidelines to avoid environmental fines and penalties (Famiyeh et al. 2018; Yalabik and Fairchild 2011).

In another comparative study, between New Zealand and Pakistan, Farrukh et al. (2023a) examined the GLSS motivators for addressing environmental sustainability in the packaging industry. In the first stage, the authors conducted a systematic literature review to identify the existing studies regarding the GLSS drivers. Then, in the second stage, the authors conducted multiple case studies in the flexible packaging firms in New Zealand and Pakistan. The study revealed interesting findings on the internal and external motivators using the in-depth insights obtained from the senior corporate managers, including the quality manager, production manager, health, safety, and environmental manager, supply chain manager, research and development manager, and maintenance manager. As such, the internal motivators were company-, operations-, and environment-driven. Company-driven motivators included top management requirements, continuous improvement, and employee pressure. Similarly, operations-driven motivators comprised cost reduction, waste reduction, and process control aspects, while environment-driven motivators included environmental responsibility, employee health and workplace safety, and sustainable consumption of resources.

Conversely, the external drivers category included state-driven, society-driven, and market-driven motivators. State-driven motivators included environmental regulations, while market-driven motivators comprised pressure from the suppliers, competitors, customers, and the company's intentions to maintain a green image in the market. The study presented similarities and differences between the two countries while adopting a GLSS approach. The similarities identified as the salient drivers among New Zealand and Pakistan were environmental responsibility, cost reduction, waste reduction, employee health and safety, and process control. In contrast, the awareness of closed-loop practices, continuous improvement, and employee pressure were revealed as the internal GLSS motivators in New Zealand flexible packaging (FP) companies compared to Pakistan companies. In addition, company image and pressure from corporate customers were revealed as significant external motivators for GLSS in both countries. On the other hand, membership organizations, NGOs, international organizations, and local development bodies were identified as the key external motivators compelling FP companies in NZ to adopt the GLSS strategy.

A study conducted in the Indian manufacturing sector identified various internal and external drivers, such as market competition, technological developments, customer requirements for GLSS, supplier pressure, and future regulatory requirements, as the key GLSS drivers (Hariyani and Mishra 2022b). In addition, market competition was determined as the highest-ranked motivator associated with the availability of a firm's resources for executing the GLSS approach. Another study, conducted in the Chinese manufacturing sector, revealed that improving the green company image is an internal GLSS driver, as customers prefer to purchase from companies producing environmentally friendly products with a clean and green image (Garza-Reyes et al. 2018a, b, c). Moreover, meeting customer requirements for eco-friendly products, complying with environmental regulations, and upholding social responsibility are key external motivators (Garza-Reyes et al. 2018a, b, c). Similarly, prior studies identified several other internal and external motivators for manufacturing firms to achieve environmental sustainability, such as reducing energy consumption by complying with regulatory requirements, developing a long-term organizational strategy, increasing energy efficiency and innovation, achieving customer satisfaction, and fostering energy knowledge (Garza-Reyes et al. 2018a, b, c; Seth et al. 2018). Table 2.1 presents the GLSS drivers for achieving environmental sustainability.

2.2.2 *Barriers to Green-Lean-Six Sigma Adoption*

GLSS barriers are identified as factors impeding organizational efforts towards integration and execution of the GLSS strategy for achieving environmental performance (Orji 2019; Shokri et al. 2022; Sreedharan et al. 2018; Yadav and Gahlot 2022b). Due to its novelty, the GLSS strategy has been executed in only a few manufacturing sectors, as the organizations face numerous barriers to its implementation (Farrukh et al. 2020; Kaswan et al. 2021). Therefore, identifying and understanding barriers inhibiting the GLSS strategy is crucial to mitigate potential future failure while executing the GLSS practices in organizations (Kaswan and Rathi 2020b; Shokri et al. 2022). The extant literature revealed different barriers hindering the organizations' successful adoption of the GLSS strategy. For example, a study by Shokri et al. (2022) investigated the barriers to GLSS through a survey questionnaire to collect data from LSS consultants and practitioners using the principal component analysis (PCA) method. The results highlighted a list of barriers, including GLSS overburdening and market challenges as the key barriers, followed by innovation and strategy deficiency, social and policy deficiency, knowledge and resource deficiency, and cultural and leadership deficiency.

Another study explored the GLSS barriers in the manufacturing industry (Kaswan et al. 2021). In phase 1, the authors identified 18 barriers to adopting GLSS through an in-depth literature review. The GLSS barriers were defined as technical and strategic issues inhibiting organizations from achieving the desired operational and environmental benefits (Aboelmaged 2011). In phase 2, a questionnaire was developed and

Table 2.1 GLSS drivers for achieving environmental sustainability

GLSS drivers			
Internal drivers	References	External drivers	References
Profit margin	Fayyaz et al. (2024), Jin et al. (2024), Shokri et al. (2022)	Customer pressure	Farrukh et al. (2023a), Gandhi et al. (2021), Jin et al. (2024), Perez-Burgoin et al. (2024)
Cost reduction	Farrukh et al. (2021a), Hariyani et al. (2025), Rahardjo et al. (2025), Sharma and Singh (2025)	Competitive advantage	Farrukh et al. (2023a)
Financial incentives	Cherrafi et al. (2016a, b)	Regulatory requirements	Hariyani et al. (2025), McDermott et al. (2024), McGlinchey et al. (2025)
Waste reduction	Farrukh et al. (2021a), Ghasemibojd et al. (2025), Shokri et al. (2022)	Influence of development organizations	Farrukh et al. (2023a)
Organization's willingness and desire to execute the GLSS approach	Kaswan and Rathi (2019)	Supplier pressure	Farrukh et al. (2023a)
Continuous improvement	Ghasemibojd et al. (2025), Hariyani et al. (2025)	Role of NGOs	Farrukh et al. (2021b)
Environmental awareness	Farrukh et al. (2020), Kaswan et al. (2021), Siegel et al. (2019)	Market competition	Farrukh et al. (2023a)
Resource circulation	Farrukh et al. (2021b)	Technological developments	Hariyani and Mishra (2022b)
Process control	Farrukh et al. (2021b)	Supplier pressure	Farrukh et al. (2023a)
Employee satisfaction	Hariyani et al. (2025), Kaswan et al. (2024)		
Process improvement	Cherrafi et al. (2016a)		
Firm image	Shokri et al. (2022)		
Employee health and workplace safety	Farrukh et al. (2023a)		
Employee pressure	Farrukh et al. (2023a)		

shared with the experts and practitioners in the manufacturing sector, including LSS black belt, LSS green belt, academics, and industry personnel. The authors used experts' opinions, including academics and practitioners, and utilized the PCA to categorize the barriers into six groups: environmental, management, knowledge, training, organizational, and continuous improvement. Moreover, the authors emphasized examining the contextual relationships among different barriers. The study highlighted that among these, management-related barriers were identified as the key issues, followed by the environmental-related barriers to GLSS implementation in the manufacturing sector. Further, the study highlighted the key barriers as a lack of customer participation, a lack of statistical and visual control over the GLSS execution, a lack of financial resources, and inadequate execution of Kaizen as the silent barriers to implementing the GLSS strategy.

In addition, another study conducted in the construction sector in a developing country examined the GLSS barriers (Hussain et al. 2019). The authors first conducted a systematic literature review (SLR) and identified 24 barriers to the GLSCP (Green-Lean-Six Sigma construction process). Subsequently, a questionnaire was developed based on the experts' opinions and responses regarding the significance of these barriers. In addition, a hierarchy of these barriers was created using a structural self-interaction matrix (SSIM). Moreover, the relationships among these barriers were also determined by using interpretive structural modelling (ISM) and Matrice d'Impacts Croisés Multiplication Appliqués à un Classement (MICMAC) analysis. Following this, an ISM-based model indicating the relationships among these barriers was developed, where the base barrier was determined as the 'unstable political environment'. The study revealed that an unstable political environment is the most critical barrier to the implementation of GLSS in the construction sector of Pakistan, followed by a lack of government support, customer involvement, and awareness of GLSS. Moreover, a lack of leadership support for adopting GLSS and a lack of funds were highlighted as the key barriers in the construction industry.

Kumar et al. (2016) studied India's automobile industry to investigate GLSS barriers. The authors identified 21 Green-Lean-Six Sigma product development barriers from an extensive literature review (Kumar et al. 2016). The authors explored the contextual relationships among these barriers based on a team of seven experts from industry and academia and used an interpretive structural modelling (ISM) technique for developing a hierarchical model of barriers. Further, the study used the MICMAC analysis to group the barriers into different categories of driving, linkage, and dependent barriers to aid practitioners in mitigating the challenges of GLSS implementation and achieving sustainable business performance. Seven barriers were found as the driver barriers, whereas nine were identified as the dependent barriers, and five barriers were considered the linkage barriers. The results indicated a lack of top management commitment as the key input barrier leading to other barriers, including lack of funds, lack of continuous improvement environment, lack of training and expertise, lack of statistical knowledge and capabilities, lack of technology, inadequate utilization of infrastructure, and competition and uncertainty.

In addition, another study identified 20 barriers regarding LSS implementation with environmental considerations (Ben Ruben et al. 2018). The authors determined the relationships among the obstacles with the help of a survey, which included a team of experts, practitioners, and academics. The authors developed an ISM model of barriers to determine the interrelationships among various obstacles. The findings revealed the most prominent barriers as a lack of finance for environmental projects, difficulty in adopting environmental practices, stringent government policies, negative attitude towards sustainability concepts, and a lack of defect monitoring analysis.

Yadav et al. (2023a, b) reviewed emerging literature to identify the challenges in GLSS execution in the higher education sector. Then, using the insights obtained from academics and experts 14 barriers were selected for further investigation. Following this, the study used the Multi-Criteria Decision-Making (MCDM) approach to rank the barriers. The findings revealed that training and education are significant challenges, followed by the continuous evaluation of sustainable development goals and inadequate organizational culture, which are the key barriers to the implementation of GLSS in the higher education industry in India.

Another study identified the barriers, such as the inadequate identification of the required areas of improvement, lack of skills and knowledge (regarding the GLSS practices) of the personnel executing the project, and inadequate understanding of the sustainability indicators, which can lead to failure of GLSS projects (Kaswan et al. 2021). In addition to the above studies, Farrukh et al. (2020) employed a systematic literature review methodology to identify a list of GLSS barriers in the manufacturing sector. The key GLSS barriers were identified as a lack of organizational awareness and information on environmental sustainability initiatives, a lack of leadership support for sustainable initiatives, a lack of financial and other resources, such as qualified and trained manpower, a lack of organizational infrastructure and continuous improvement opportunities, and a lack of human resource management.

Similarly, Hariyani and Mishra (2022a) conducted a literature review to determine the GLSS barriers using the Google Scholar and Web of Science databases and identified 31 barriers. The findings highlighted weak regulatory framework, lack of government support, lack of consumer knowledge, low public awareness, complex design of the environmentally friendly products, lack of leadership support, lack of training, lack of technical resources, and lack of supplier involvement as the key hurdles in GLSS implementation. In addition, the extant literature revealed a causal relationship among barriers: for example, a study conducted by Kaswan et al. (2021) revealed the top barriers related to environmental, management, and organizational factors due to their causal relationships with other barriers.

Building on the literature review, the GLSS barriers are classified as organizational, economic, technical, individual, human resource, and socio-cultural. Among these, prior studies highlighted different organizational barriers such as resistance to change, lack of top management commitment, lack of resources to execute GLSS practices, lack of adequate sustainability policy, lack of motivation to implement GLSS strategy, inadequate policies, and lack of environmental regulations (Belhadi et al. 2020; Hariyani et al. 2023; Kaswan and Rathi 2020b; Mohan et al. 2023; Yadav et al. 2023a, b). However, these challenges can be addressed by developing the LSS maturity level in organizations during the transformation phase (Shokri et al. 2022). On the other hand, organizational size and characteristics were also recognized as key organizational factors in GLSS implementation (Trianni et al. 2016). De Freitas and Costa (2017) identified the trade-off between environmental and economic performance as one of the significant organizational barriers of the GLSS strategy. Similarly, another study by Cherrafi et al. (2016a) noted the conflict between environmental and economic sustainability, including the customer and quality-oriented LSS as a key organization-level inhibitor to the GLSS execution. Along with the above, organizational barriers such as a lack of synergy between the organization's objective(s), the continuous improvement approach, and fear of failure among employees can also prohibit the execution of GLSS. Accordingly, practitioners must identify and recognize the types of barriers and then adopt adequate solutions (Kaswan et al. 2021).

The organizational barriers also include an inadequate regulatory framework, poor quality of raw materials, and inadequate organizational strategies and plans for GLSS execution. Other organizational barriers were inefficient transportation, a lack of material handling, and a lack of top management support (Yadav and Gahlot 2022a). Further, inadequate infrastructure and production floor utilization, lack of time management, and lack of organizational attention towards Kaizen initiatives are significant barriers impeding GLSS execution (Yadav and Gahlot 2022a). In addition, cultural issues and differences inhibiting mutual learning and promoting green culture in the manufacturing firms are also related to the organizational barriers (Kaswan et al. 2021). Moreover, a lack of organizational capability towards continuous improvement thinking, learning, and implementation of re-engineering methods and sustainable practices also hinders the GLSS execution.

In terms of economic barriers, financial constraints also impede the successful deployment of the GLSS strategy, as sufficient funds are required to execute the sustainable practices, such as conducting life cycle analysis; implementing an environmental management system (ISO 14001); developing sustainable raw materials, such as biodegradable and bio-based materials; and deploying a recycling plant, such as a waste recycling or water recycling plant (Bening et al. 2021; Farrukh and Sajjad 2023; Kaswan et al. 2021). In addition, the lack of financial resources can be attributed to the organization's financial constraints, as well as the unavailability of loans from the government to adopt sustainable practices (Kaswan et al. 2021). Along with the fear of failure of the GLSS strategy, the uncertainty of return on investment from the GLSS practices is another barrier hindering the execution of this strategy (Hussain et al. 2019). Furthermore, prior studies also highlighted the human resource-related barriers, such as a lack of human skills and knowledge, inadequate selection of GLSS tools, insufficient use of human resources, a lack of understanding among employees, and a lack of experience among personnel handling GLSS projects, which can impede the successful deployment of the GLSS practices (Kaswan et al. 2021).

On the other hand, technical barriers include the lack of digital technologies, the lack of digital training and knowledge, and internal resistance within the firm to adopt digitalization (Yadav and Gahlot 2022a). Moreover, inadequate identification of process parameters, lack of availability of accurate information and data, and lack of identification of green and lean metrics and relationships are the factors impeding adoption of digital initiatives (Kaswan et al. 2021). Table 2.2 highlights several barriers to GLSS adoption to achieve environmental sustainability.

2.2.3 Green-Lean-Six Sigma Practices and Tools

The extant literature highlighted a large number of studies investigating the GLSS practices and tools to address the issues of operational and environmental waste reduction. Numerous researchers have investigated a diverse range of GLSS tools for achieving sustainable performance (Iqbal et al. 2018; Jamil et al. 2020; Mohammad et al. 2024; Tsarouhas and Sidiropoulou 2024; Yadav and Gahlot 2022a). In this regard, researchers classified the GLSS practices and tools into green, lean, and Six Sigma practices (Garza-Reyes et al. 2018a, b, c; Yadav and Gahlot 2022b). Among the green tools, prior studies examined the application of various green practices such as 3R, LCA, EMS, eco-labelling, ISO 14001, and DFE (Ahmed et al. 2021; Walter et al. 2021). In addition, various studies examined the application of different lean tools such as 5S, VSM, TPM, Kaizen, standardization, SMED, Poka-yoke, etc., to address the environmental performance of organizations (Garza-Reyes et al. 2018a, b, c; Mohammad et al. 2024). On the other hand, several Six Sigma tools, such as DMAIC, cause-and-effect analysis, SIPOC, Pareto diagram, SPC, box plot, regression analysis, and DMDAV, are used to address both the operational and environmental performance of organizations (Rathi et al. 2022; Tsarouhas and Sidiropoulou 2024; Walter et al. 2021). The following Table 2.3 includes a brief overview of different GLSS tools.

A study investigated the application of various GLSS tools in the plastic packaging industry (Farrukh et al. 2022a). The authors conducted a multiple case study approach and interviewed senior corporate managers in two large manufacturing companies. The findings highlighted that the 3R strategy was used, such as recycling practices, which were followed in the case companies to conserve resources. Along with recycling, companies also encourage the reuse of resources, such as reusing inks from printed packaging materials. In addition, the packaging material was re-utilized in the job setting process in the printing area. Further, a solvent recovery plant was also installed to recover the used solvents from the printing and lamination areas, which can be reused to clean ink drums. On the other hand, various energy management practices were also followed in the companies, such as solar panels for producing renewable energy, ISO 50001 energy management system certification, energy-saving lights, sunlight, and variable frequency devices and inverters to balance the energy consumption with the requirements of the machines.

From a lean perspective, the study also highlighted the application of various practices in achieving operational and environmental performance, such as SMED for reducing setup time, materials (solvents, printing films, and inks), energy, and

Table 2.2 Barriers to GLSS adoption

Barriers to green-lean six sigma		References
Organizational barriers	– Resistance to change – Lack of top management commitment – Lack of an adequate sustainability policy – Lack of motivation for implementing the GLSS strategy – Inadequate policies – Lack of environmental regulations – Organizational size and characteristics – Trade-off between environmental and economic performance – Lack of synergy between the organization's objective(s) and the continuous improvement approach – Fear of failure among employees – Poor quality of raw material – Inefficient transportation – Lack of material handling	Belhadi et al. (2020), Hariyani et al. (2023), Kaswan and Rathi (2020b), Mohan et al. (2023) Yadav et al. (2023a, b)
Economic barriers	– Financial constraints – Insufficient funds to conduct life cycle analysis – Lack of financial resources for implementing an environmental management system – Lack of financial resources for developing sustainable raw materials – Uncertainty of return on investment from the GLSS	Bening et al. (2021), Farrukh and Sajjad (2023), Kaswan et al. (2021)
Technical barriers	– Lack of digital technologies – Lack of digital training and knowledge – Internal resistance within the firm to adoption of digitalization – Inadequate identification of process parameters – Lack of availability of accurate information and data – Lack of identification of green and lean metrics and their relationships	Yadav and Gahlot (2022a)
Individual, human resource, and socio-cultural barriers	– Lack of human skills and knowledge – Inadequate selection of GLSS tools – Insufficient use of human resources – Lack of employee understanding – Lack of experience of personnel handling GLSS projects	Kaswan et al. (2021)

Table 2.3 Overview of GLSS tools

GLSS tools	Description	References
5S	5S methodology involves separating, setting in order, shining, standardizing, and sustaining for achieving operational and environmental sustainability through waste collection and sorting, reducing emissions, optimizing space for workplace activities, and reducing leakage, fumes, and solvents	Bouazza et al. (2021), Brown et al. (n.d.), Setiawan et al. (2021)
VSM	VSM is used for highlighting, mapping, and analysing the operational and environmental waste of various processes in product manufacturing in organizations. As such, the VSM can identify resource consumption, delays, air emissions, and energy consumption	Andreazza de Freitas et al. (2025), Horsthofer-Rauch (2025), Larsson et al. (2025), Nguyen and Sharmak (2022)
Cellular manufacturing	Cellular manufacturing is focused on grouping the equipment, machines, and staff dedicated to similar products in a single U-shaped cell in manufacturing organizations, which helps reduce worker fatigue, unnecessary transportation of products and materials, fuel consumption, and air emissions	Aadithya et al. (2023), Bouazza et al. (2021), Sharma et al. (2022)
SMED	SMED helps reduce machine setup time through efficient management and organization of activities necessary for changing machine parts, which results in energy conservation, resource efficiency, and reduced lead time	Ebrahimi et al. (2023) Lino Moreno et al. (2025), Rahardjo et al. (2023)
LCA	Life cycle analysis helps determine the environmental impact of products and processes at different stages, including cradle to gate, cradle to grave, and cradle to cradle	Fernando et al. (2025) Khamidov et al. (2023), Tan et al. (2023)
3R	3R practices include reducing, reusing, and recycling products and materials, enabling firms to achieve sustainability and circular economy objectives	Abd Rahim et al. (2025), Iliná et al. (2023), Mohammed et al. (2021)
EMS	An environmental management system, such as ISO 14001, facilitates organizations' reduction of the environmental impact of their products and processes while addressing stakeholder needs and requirements regarding eco-friendly products and environmental safety and protection	Ahmed et al. (2024), Bhateria (2024), Ong et al. (2023), Riaz et al. (2022), Wessels and Ross (2024), Yang et al. (2024)
DMAIC	DMAIC consists of five stages (define, measure, analyse, improve, and control) which help organizations identify and control defects and operational and environmental waste generated from their processes	Dahab et al. (2024), Fayyaz et al. (2025), Tibebu and Henry (2025), Wang et al. (2025)

(continued)

Table 2.3 (continued)

GLSS tools	Description	References
SPC	SPC helps control processes and reduce variation, leading to waste minimization, through the application of different charts, such as the X-bar R chart, U chart, P chart, and X-bar S chart	Adnan and Nasir (2025), da Silva and Borges (2025), Nkansah (2025), Swarnakar and Khalfan (2025)
TPM	TPM focuses on reducing machine failure and stoppages through regular maintenance, which can reduce oil leakage, air emissions such as volatile organic compounds, dust, and fumes	Crosby and Badurdeen (2022), Hallioui et al. (2023), Morales-García et al. (2021), Samadhiya and Agrawal (2024), Samadhiya et al. (2023)
JIT	JIT refers to optimal inventory management, which ensures that only the required material is available at the necessary time in the workplace. It focuses on waste reduction through minimizing unnecessary inventory, which further reduces environmental waste such as excessive energy consumption, air emissions, and obsolete inventory	Agyabeng-Mensah et al. (2021), Lara et al. (2022), Rashid et al. (2025), Singhal et al. (2025)
Cause-and-effect analysis	Cause-and-effect analysis focuses on identifying the root causes of operational and environmental waste, considering the potential reasons associated with man, machine, method, measurement, and environment	Cucurachi and Suh (2017), Nababan et al. (2018), Ren et al. (2013), Wałach (2020)
Pareto diagram	A Pareto diagram is a problem-solving tool. It prioritizes problems according to the 80–20 principle, which emphasizes that a few vital problems lead to the main effect/issue	Ait Hammou and Oulfarsi (2022), Hassan et al. (2025), Kurnia et al. (2022), Paliska et al. (2007)

air emissions. In addition, ECRS (eliminate-combine-rearrange-simplify) was identified to reduce the operational and environmental waste, such as excessive energy and process. Similarly, Kaizen was used to optimize the printing machine routes to save energy consumption. Moreover, 5S was deployed to optimize energy consumption by removing unnecessary machine motors. In addition, total productive maintenance was highlighted as one of the key lean tools addressing operational and environmental waste, such as waiting time, process waste, machine breakdown time, energy consumption, liquid discharge, and noise pollution. Along with the green and lean tools, the study revealed the benefits of using Six Sigma tools, such as DMAIC for process waste reduction and energy consumption, cause-and-effect analysis for defect detection in the packaging to reduce food waste, and control charts for plastic waste reduction by controlling the grammage variation. In addition, the Pareto diagram and cause-and-effect analysis help determine the reasons behind machine downtime.

Another study conducted in the substrate manufacturing sector conducted action-based research using a DMAIC methodology (Gholami et al. 2020). In the define stage, the authors identified the research's scope and objectives, such as identifying and measuring environmental wastes related to the electroless nickel plating (ENP) process. The environmental metrics were identified, such as GHG emissions, carbon footprint, process water consumption, water eutrophication, energy consumption, noise level, raw material usage, and chemical consumption. In the measure phase, the authors selected the relevant metrics for measuring key process parameters to align with the project objectives. In particular, the measure stage included two sub-stages, such as establishing the environmental metrics, including the process water consumption, chemical consumption, and energy consumption, as the relevant environmental metrics to enhance the overall performance of the manufacturing process. Then, the authors developed the current state environmental-VSM (E-VSM) using the selected metrics.

The authors identified the difference between the actual and target performance, analysed the results, and proposed solutions for the ENP process (Gholami et al. 2020). In this context, the study utilized the current state of E-VSM to evaluate the environmental issues, such as the need to reduce chemical and energy consumption. Following this, the authors developed strategies to reduce environmental waste by reducing the phosphorus content in the ENP chemicals, which can reduce overall chemical consumption, optimize the process, and reduce the bleed-off volume while ensuring the quality and effectiveness of the plating. The results highlighted a 28% reduction in chemical consumption, further reducing the material costs (Gholami et al. 2020).

Meanwhile, the Pareto diagram was used to analyse the electricity distribution, which highlighted the highest energy consumption (94659 K) in the post-plating oven (PPO) processes among other processes, such as the plating process (72572 K). The authors proposed altering the operating controller and sensor to reduce energy consumption, which could result in a 21% energy reduction and achieve the cost-saving objectives. Using the proposed chemical and energy reduction strategies, the future-state E-VSM for the ENP process was developed in the improvement stage to increase environmental performance. Finally, in the control stage, the standard operating procedures for continuously monitoring the deployed improvement strategies were established to ensure a uniform process implementation and minimum variation in the environmental parameters. From this perspective, statistical process control tools, such as the control charts (Xbar-R and boxplot), were used via the Minitab software to monitor the energy and chemical consumption. Moreover, the best practices and improvement measures regarding operational and environmental waste were recorded and documented as part of the SOPs to be followed by all employees.

Similarly, another study conducted a comprehensive literature review and identified that among the most common GLSS tools, 5S as a lean tool was utilized around

Table 2.4 GLSS tools for addressing environmental sustainability

Green-lean-six sigma tools		References
Green tools	LCA, EMS, 3R, 6R, 9R, EMS, DFE, ISO 50001, ISO 450001, carbon calculators, ecolabelling, energy audits, Enviro-mark, renewable energy sources, E-WOF, hazardous substance inventory tools	Almomani et al. (2025), Farrukh et al. (2022b, 2023a), Horodytska et al. (2020), Kennelly et al. (2019), McDermott et al. (2024), Tasdemir et al. (2019)
Lean tools	5S, Kaizen, Poka-yoke, FMEA, lean Mudas, TPM, JIT, 5 Whys, cellular manufacturing, cost–benefit analysis, Gemba walk, Pareto diagram, process mapping, motion study, VSM, plan-do-check-act cycle, work standardization	Ghosh et al. (2023, 2025), Jonda et al. (2023), Jovanović et al. (2017), Lizarelli et al. (2024, 2025), McDermott et al. (2024)
Six sigma tools	SPC. SIPOC, QFD, ANOVA, DMAIC, DMADV, process capability analysis, project charter, critical to quality, design of experiment	Al Abri (2023), Fernando et al. (2024), Kurnia et al. (2022), Rodriguez Delgadillo et al. (2022), Santoso et al. (2024), Shrivastava and Mishra (2025) Siwiec and Pacana (2025)

80%, SIPOC diagram as a Six Sigma tool was implemented around 86%, and E-VSM as a green tool was used 80% of the time in several organizations (Yadav and Gahlot 2022a). The study also revealed that SMEs prefer using lean tools to address environmental issues. Table 2.4 demonstrates the GLSS tools used by organizations to address environmental sustainability.

2.2.4 *Critical Success Factors (CSFs) of Green-Lean-Six Sigma Implementation*

Along with implementing different tools, the success of the GLSS strategy is premised on the critical factors, which refer to the areas requiring significant attention for the successful deployment of this approach (Yadav and Gahlot 2022a). Numerous studies identified diverse types of CSFs for the GLSS implementation in organizations (Farrukh et al. 2019; Ghadimi et al. 2020; Mishra 2018; Rathi et al. 2021; Seth and Rehman 2022; Sreedharan et al. 2018). For example, one study highlighted government support and incentives, leadership support, and availability of financial resources to successfully deploy the GLSS approach for improving environmental performance, enhancing firms' capabilities, and achieving the maturity level of GLSS practices (Kramat Hussain et al. 2022a, b). In addition, Walter et al. (2021) highlighted various CSFs of a GLSS strategy using the systematic literature review methodology, including employee knowledge and awareness of GLSS, leadership commitment and communication, organizational culture to foster continuous improvement and innovative culture, employee involvement and participation, teamwork, and appropriate project selection.

Building on the existing literature, the CSFs of a GLSS strategy can be classified as strategic, human resource, operational, technical, and stakeholders. For example, strategic factors include the top management support, commitment, and involvement as key factors of the GLSS (Cherrafi et al. 2016b; Yadav and Gahlot 2022a). As far as the role of top management is concerned, their support is one of the key strategic factors for executing new practices and adopting novel strategies in an organization (Kaswan et al. 2021; Kattumuri 2018). Hence, mitigating barriers to the GLSS implementation requires top management support, responsibility, and commitment (Cherrafi et al. 2016a). For example, the top management commitment towards adopting clean technologies (a green practice) is vital for incorporating sustainability in an organization (Kaswan et al. 2021; Kattumuri 2018).

Moreover, top management support is a catalyst for developing a culture of continuous learning, improving employees' confidence, and building trust among the organization's members (Kaswan et al. 2021). Along with the above, their vision, involvement, and patronage can play a significant role in executing the GLSS approach in firms' operations (Kaswan et al. 2021; Kattumuri 2018). Further, the top management's efforts to integrate sustainability with organizational objectives not only demonstrate their commitment towards executing an environmentally-friendly strategy (such as GLSS) but also promote change and mitigate the fear of adopting new practices (Kaswan et al. 2021; Kattumuri 2018). A study emphasized that the top management's awareness of the GLSS strategy can be increased by engaging an organization's leadership in brainstorming activities, demonstrating the advantages of the GLSS tools, and providing advisory and consultancy services (Hussain et al. 2019).

Human resources in an organization play an effective role in developing the organizational culture and implementing organizational strategies. With the emergence of the green manufacturing paradigm, firms have also directed their attention towards developing green activities in the workforce by focusing on employee involvement and commitment to achieving environmental objectives. Furthermore, organizations need to create a conducive work environment where employees are empowered and can easily interact with top management to share their suggestions regarding environmental improvements (Masri and Jaaron 2017). Human resource-related factors focus on employee involvement and participation in executing the GLSS strategy (Yadav and Gahlot 2022a).

Similarly, employees must be aware of their roles and responsibilities before starting the improvement projects, such as GLSS projects (Alhuraish et al. 2017). Furthermore, employee commitment plays a significant role in these projects' success (Soti et al. 2010; Tjahjono et al. 2010). Employee empowerment, teamwork, and high-quality communication among team members are essential for executing an operational excellence strategy (Tlapa et al. 2016). Additionally, the team members should have statistical and project management skills to successfully implement these projects (Alhuraish et al. 2017).

In addition, the employees' knowledge regarding the GLSS strategy can contribute to achieving the organization's goals and objectives of sustainable performance. Employee awareness of utilizing adequate GLSS tools according to the requirements of environmental projects is another vital factor (Farrukh et al. 2019). Similarly, employees' knowledge of GLSS practices, training, teamwork, and employees' competencies play a significant role in the success of a GLSS strategy (Cherrafi et al. 2016b; Farrukh et al. 2019, 2023b; Mishra 2018; Walter et al. 2021). As such, the employees' training on GLSS techniques and concepts can enhance their knowledge, skills, and development for successfully executing this strategy (Farrukh et al. 2019; Hussain et al. 2019). Further, the GLSS projects are more likely to be successful by developing cross-functional teams and through collaboration and joint ventures with industry and academia. Initiating employee training programmes and workshops on the GLSS techniques can contribute to effectively utilizing resources for achieving sustainability (Ahmed and Mathrani 2023; Letchumanan et al. 2022). Along with the above, motivating and encouraging employees to learn new skills and sustainable practices can also promote GLSS adoption in organizations (Ahmed and Mathrani 2023). Further, the coordination and communication among employees can be enhanced by fostering information sharing through adequate communication technology and networking (Farrukh et al. 2019).

On the other hand, the operational and technical factors include the measurement and relevant metrics for achieving sustainable performance from the GLSS implementation (Yadav and Gahlot 2022a; Cherrafi et al. 2017). Among the operational factors, appropriate selection of the GLSS project with a clear focus on the sustainable objectives is one of the significant factors. From this perspective, appropriate selection depends on the personnel's knowledge and experience, including understanding the GLSS tools, project objectives, and sustainable outcomes from the selected projects (Kaswan et al. 2021; Kattumuri 2018). Further, operational factors such as measurement system, data accuracy, suitable performance metrics, and feedback system are also considered. Moreover, a supportive organizational culture and reward system can also enhance the successful deployment of the GLSS approach. In this regard, a study highlighted that the reward system in the form of recognition and financial incentives for middle- and operational-level employees is identified as a critical success factor in executing GLSS. Further, benchmarking the sustainable practices and initiating the certification system of the GLSS practices, similar to the Six Sigma black belt and ISO 14001, can promote the execution of GLSS (Ahmed and Mathrani 2023; Hussain et al. 2019).

Additionally, technological advancements such as innovative technologies and information communication technologies (ICT) can play a significant role in executing the GLSS practices, which can enhance better decision-making and information sharing, leading to efficient management of the GLSS projects (Hussain et al. 2019; Letchumanan et al. 2022). Along with the above, integrating sustainable practices such as JIT with technologies, including big data, can also improve the raw material inventory and logistics management. Moreover, visual management and statistical software for managing quality, controlling manufacturing operations, and measuring sustainability indicators (such as environmental impact) can also aid the GLSS implementation (Hussain et al. 2019).

The literature also highlighted the stakeholder-related CSFs for the successful deployment of a GLSS strategy (Ahmed and Mathrani 2023; Farrukh et al. 2023b; Mohan et al. 2024b; Shokri and Li 2020; Yadav and Gahlot 2022a; Yadav et al. 2023a, b). In this perspective, the stakeholders such as government, suppliers, customers, recyclers, waste collectors, policy makers, and regulatory bodies can also play a significant role in promoting the benefits of adopting sustainable strategies such as GLSS (Kaswan et al. 2021; Mohan et al. 2024b). Despite organizations' willingness to implement the GLSS approach, the level of adoption is low. This is attributed to stakeholders' lack of awareness of the potential sustainable benefits of GLSS in the form of cost and emission reduction, and optimal utilization of natural resources. In addition, campaigns and programmes can enhance stakeholders' awareness and knowledge of sustainability. For instance, increasing customer awareness of the GLSS advantages can enhance their demand for eco-friendly products, which will help organizations minimize the fear of failure among employees (Kramat Hussain et al. 2022a, b; Hussain et al. 2019).

Further, stakeholders' awareness can be increased through consultation and brainstorming sessions regarding the benefits of the GLSS, as the stakeholders, including manufacturers and suppliers, are concerned about the operational cost regarding the adoption of sustainable technologies, cooperation, research and development, and circular economy practices, including recycling, durability, and reusability of materials and parts. A study highlighted that effective communication and information sharing with stakeholders regarding operations at different stages of the construction process can also play a significant role in the GLSS implementation (Hussain et al. 2019). Conversely, stakeholders' understanding of GLSS strategy can improve their raw material quality and intra-organizational collaboration.

Accordingly, adequate support from the stakeholders (in particular, from the government), can motivate the organizations towards initiating sustainable practices such as the GLSS (Farrukh et al. 2023b; Kaswan et al. 2021; Mohan et al. 2024b). A review of the literature revealed that government support and adequate policies have a positive impact on the adoption of eco-friendly practices. For example, a study highlighted the successful execution of a Green-Lean project in the manufacturing industry as a result of collaboration with the Washington state Department of Ecology, which supported various manufacturing firms to improve sustainable performance (Kaswan et al. 2021). The government can provide loans at low interest rates to industries which can promote GLSS in organizations. The government support can also overcome the barrier of uncertainty of return from the GLSS strategy, with adequate support (Kramat Hussain et al. 2022a, b; Hussain et al. 2019). The government can also encourage foreign investment in manufacturing industries, such as the construction sector, to execute sustainable practices such as GLSS. A study conducted in the construction sector in Pakistan emphasized GLSS execution through adequate policies, stringent regulations, and government support in the form of subsidies, tax concessions, and rewards (Hussain et al. 2019).

In addition, the suppliers can also play an important role in successfully deploying the GLSS approach by improving and utilizing their capabilities. For example, the suppliers' capabilities regarding environmental concerns can be analysed in several ways, such as environmental management systems (e.g., ISO 14001), environmental policy, and certifications (Farrukh et al. 2022a). Additionally, manufacturing organizations help their suppliers become green by providing appropriate training, information, and facilitation (Asiaei et al. 2023; Farrukh et al. 2022a). For example, workshops and seminars on environmental aspects enhance their knowledge and information. Similarly, organizations can develop environmental teams who can visit suppliers' sites and provide them with technical support and financial assistance where needed in their processes (Lee et al. 2015). Supplier development positively affects a manufacturing firm's inventories, lead times, and customer satisfaction. Additionally, maintaining supplier relations includes suppliers' awareness regarding the buyer firm's manufacturing requirements, supplier participation, supplier evaluation, and supplier commitment (Garza-Reyes et al. 2018a, b, c). Table 2.5 presents the strategic, human resource, operational, technical, and stakeholder-related CSFs of the GLSS strategy for achieving environmental sustainability in organizations.

Table 2.5 CSFs of the GLSS strategy for achieving environmental sustainability

CSFs of GLSS		References
Strategic factors	Top management commitment	Kramat Hussain et al. (2022a, b), Ishak et al. (2024) Kaswan and Rathi (2019, 2020b), Kaswan et al. (2021) Mohan et al. (2024b, 2025b), Perez-Burgoin et al. (2024), Rathi et al. (2023), Shokri et al. (2022), Yadav et al. (2023a, b)
	Adequate funds	Hudnurkar et al. (2019), Kaswan et al. (2021), Parmar and Desai (2020a, b), R et al. (2018), Thomas and Khanduja (2021)
	Project leadership	Ishak et al. (2024)
	Linking GLSS to business objectives	Kaswan and Rathi (2019, 2020a), Rathi et al. (2023), Yadav et al. (2023a, b)
	Adequate communication and cooperation	Kaswan et al. (2021)
	Resource allocation	Ishak et al. (2024), Mohan et al. (2025b), Rathi et al. (2023), Shokri et al. (2022), Yadav et al. (2023a, b)
	Integrated approach	Mohan et al. (2025a), Rathi et al. (2023)
	Organization culture	Farrukh et al. (2020), Yadav et al. (2023a, b)

(continued)

Table 2.5 (continued)

CSFs of GLSS		References
	Organizational capability and quality maturity level for GLSS operations	Hussain et al. (2022a, b)
	Organizational readiness for GLSS adoption	Kaswan and Rathi (2019, 2020b), Rathi et al. (2023), Yadav et al. (2023a, b)
Operational and technical factors	Project management	Shokri et al. (2022), Yadav et al. (2023a, b)
	Performance measurement	Kaswan and Rathi (2019), Perez-Burgoin et al. (2024)
	Precise data collection	Ishak et al. (2024), Mohan et al. (2025b)
	Adequate feedback system	Kaswan and Rathi (2019)
	Integrating sustainable measures	Rathi et al. (2023)
	Organization engagement	Perez-Burgoin et al. (2024)
	Employee involvement	Farrukh et al. (2020), Mohan et al. (2025a)
	Real-time data collection	Rathi et al. (2023)
	Project manager's support	Shokri et al. (2022)
	Integrated environmental sustainability framework	Shokri et al. (2022)
	Corrective actions	Boumsisse et al. (2025), Cherrafi et al. (2017) Rathi et al. (2023), Tsarouhas and Sidiropoulou (2024)
	Adequate utilization of tools and techniques	Mohan et al. (2024a)
	Continuous assessment of sustainable development goals	Yadav et al. (2023a, b)
	Adequate measurement and metrics	Farrukh et al. (2020), Mohan et al. (2024a), Yadav et al. (2023a, b)

(continued)

Table 2.5 (continued)

CSFs of GLSS		References
Human resource-related factors	Employee motivation	Antony et al. (2019, 2020), Mohan et al. (2025a), Parmar and Desai (2020a), Sony et al. (2020), Swarnakar et al. (2020)
	Employee training	Farrukh et al. (2020), Ishak et al. (2024), Rathi et al. (2023), Yadav et al. (2023a, b)
	GLSS expertise	Kaswan et al. (2021)
	Employee education	Duarte and Cruz-Machado (2014), Farrukh et al. (2020, 2023b) Singh Sangwan et al. (2014), Yadav et al. (2020)
	Reward system	Gaikwad and Sunnapwar (2020), Hussain et al. (2019), Mohan et al. (2025a), Silva et al. (2019), Sreedharan et al. (2018)
	Employee empowerment	Farrukh et al. (2023b), Hariyani et al. (2025), Kumar and Rodrigues (2020), Panwar et al. (2017), Singh et al. (2021), Sodhi et al. (2020), Thanki et al. (2016)
	Employee engagement	Antony et al. (2018), Caiado et al. (2018), Deepak and Dharmendra (2025), Hariyani et al. (2025), Klein et al. (2024), Rathi et al. (2023), Siegel et al. (2019), Walter et al. (2021)
Stakeholder-related factors	Strong relationship with customers and suppliers	Yadav et al. (2023a, b)
	Government policies	Kaswan et al. (2021)
	Government support	Hussain et al. (2022a, b)
	Role of recyclers and waste collectors	Farrukh (2025), Farrukh et al. (2023b)

2.3 Summary

This chapter discusses the theoretical constructs of the Green-Lean-Six Sigma (GLSS) approach for improving environmental sustainability including drivers, barriers, tools, and CSFs. The chapter highlights various internal and external motivators behind the GLSS adoption in organizations. Following this, challenges to the GLSS implementation are discussed under different categories of organizational, economic, technical, and individual, human resource, and socio-cultural barriers. Next, various GLSS tools for enhancing environmental performance are presented including 5S, VSM, SMED, LCA, 3R, EMS, DMAIC, SPC, TPM, JIT, CE, and Pareto diagram. Finally, the CSFs of the GLSS strategy are explained under the strategic, operational and technical, human resource, and stakeholder-related factors.

References

Aadithya, B. G., Asokan, P., & Vinodh, S. (2023). Lean manufacturing in fabrication industry: literature review and framework proposal. *International Journal of Quality & Reliability Management*, *40*(6), 1485–1517.

Abd Rahim, M. A., Ghazali, S. A., Mohd Zahid, A. Z., Dzulkafli, N. F., & Kassim, J. (2025). Evaluating 3R waste management practice for construction and demolition waste among contractors in Malaysia. *Journal of Sustainable Civil Engineering & Technology (JSCET)*, *4*(2), 78–89.

Aboelmaged, M. G. (2011). Reconstructing Six Sigma barriers in manufacturing and service organizations: The effects of organizational parameters. *International Journal of Quality & Reliability Management, 28*(5), 519–541.

Adnan, M. H., & Nasir, N. (2025). Monitoring Air Quality Trends for Industrial and Urban Areas. *Bioresources and Environment*, *3*(2), 28–43.

Agyabeng-Mensah, Y., Afum, E., Agnikpe, C., Cai, J., Ahenkorah, E., & Dacosta, E. (2021). Exploring the mediating influences of total quality management and just in time between green supply chain practices and performance. *Journal of Manufacturing Technology Management*, *32*(1), 156–175.

Ahmed, A., & Mathrani, S. (2023). Critical success factors for a combined lean and ISO 14001 implementation in the manufacturing industry: a systematic literature review. *The TQM Journal*, *36*(7), 2071–2106. https://doi.org/10.1108/tqm-12-2022-0347

Ahmed, A., Mathrani, S., & Jayamaha, N. (2021). An integrated lean and ISO 14001 framework for environmental performance: an assessment of New Zealand meat industry. *International Journal of Lean Six Sigma*, *ahead-of-print*(ahead-of-print). https://doi.org/10.1108/ijlss-05-2021-0100

Ahmed, A., Mathrani, S., & Jayamaha, N. (2024). An integrated lean and ISO 14001 framework for environmental performance: an assessment of New Zealand meat industry. *International Journal of Lean Six Sigma*, *15*(3), 567–587.

Ait Hammou, I., & Oulfarsi, S. (2022). Lean Six Sigma tools and sustainable performance measurments: A Review.

Al Abri, A. M. (2023). *Identifying Potential Solutions of Grit Blasting Waste Mitigation Using Six-Sigma for Oil and Gas Industries*. Sultan Qaboos University (Oman).

Albliwi, S. A., Antony, J., & Lim, S. A. h. (2015). A systematic review of Lean Six Sigma for the manufacturing industry. *Business process Management journal*, *21*(3), 665–691. https://doi.org/10.1108/bpmj-03-2014-0019

Alhuraish, I., Robledo, C., & Kobi, A. (2017). A comparative exploration of lean manufacturing and six sigma in terms of their critical success factors. *J. Clean. Prod., 164*, 325–337. https://doi.org/10.1016/j.jclepro.2017.06.146

Almomani, M. M., Mayyas, Y. O., Alomari, O. H., Tashtoush, G. M., Cherdkeattikul, S., & Akafuah, N. K. (2025). Augmenting energy efficiency in automotive paint ovens: a review of future prospects and potential for lean, six sigma, AI, and IoT integration. *Management of Environmental Quality: An International Journal*. https://doi.org/10.1108/meq-03-2025-0164

Andreazza de Freitas, G., Hernandes de Paula e Silva, M., & Aparecido Lopes Silva, D. (2025). Overall lean and green effectiveness based on the environmentally sustainable value stream mapping adapted to agribusiness. *International Journal of Lean Six Sigma*, *16*(1), 25–53.

Antony, J., Gupta, S., Sunder M, V., & Gijo, E. V. (2018). Ten commandments of Lean Six Sigma: a practitioners' perspective. *International Journal of Productivity and Performance Management*, *67*(6), 1033–1044. https://doi.org/10.1108/ijppm-07-2017-0170

Antony, J., Lizarelli, F. L., Fernandes, M. M., Dempsey, M., Brennan, A., & McFarlane, J. (2019). A study into the reasons for process improvement project failures: results from a pilot survey. *International Journal of Quality & Reliability Management*, *36*(10), 1699–1720. https://doi.org/10.1108/ijqrm-03-2019-0093

Antony, J., Lizarelli, F. L., & Machado Fernandes, M. (2020). A Global Study Into the Reasons for Lean Six Sigma Project Failures: Key Findings and Directions for Further Research. *IEEE Transactions on Engineering Management*, 1–16. https://doi.org/10.1109/tem.2020.3009935

Asiaei, K., O'Connor, N. G., Barani, O., & Joshi, M. (2023). Green intellectual capital and ambidextrous green innovation: The impact on environmental performance. *Business Strategy and the Environment, 32*(1), 369–386.

Belhadi, A., Kamble, S. S., Zkik, K., Cherrafi, A., & Touriki, F. E. (2020). The integrated effect of Big Data Analytics, Lean Six Sigma and Green Manufacturing on the environmental performance of manufacturing companies: The case of North Africa. *Journal of Cleaner Production, 252.* https://doi.org/10.1016/j.jclepro.2019.119903

Ben Rubin, R., Vinodh, S., & Asokan, P. (2018). ISM and Fuzzy MICMAC application for analysis of Lean Six Sigma barriers with environmental considerations. *International Journal of Lean Six Sigma, 9*(1), 64–90. https://doi.org/10.1108/ijlss-11-2016-0071

Bening, C. R., Pruess, J. T., & Blum, N. U. (2021). Towards a circular plastics economy: Interacting barriers and contested solutions for flexible packaging recycling. *Journal of Cleaner Production, 302.* https://doi.org/10.1016/j.jclepro.2021.126966

Bhateria, R. (2024). Environmental Management System (EMS). In *Environmental Impact Assessment: A Journey to Sustainable Development* (pp. 101–124). Springer.

Bouazza, Y., Lajjam, A., & Dkhissi, B. (2021). The impact of lean manufacturing on enviromental performance in Moroccan automotive industry. *Management Systems in Production Engineering.*

Boumsisse, I., Benhadou, M., & Haddout, A. (2025). Optimizing Green Lean Six Sigma using Industry 5.0 technologies. *Cleaner Waste Systems, 10.* https://doi.org/10.1016/j.clwas.2025.100234

Brown, D. M., Johnson, S. L., & Thompson, M. E. (n.d.). Exploring the Impact of 5S Implementation in Industrial Organizations: A Review.

Caiado, R., Nascimento, D., Quelhas, O., Tortorella, G., & Rangel, L. (2018). Towards Sustainability through Green, Lean and Six Sigma Integration at Service Industry: Review and Framework. *Technological and Economic Development of Economy, 24*(4), 1659–1678. https://doi.org/10.3846/tede.2018.3119

Cherrafi, Elfezazi, S., Govindan, K., Garza-Reyes, J. A., Benhida, K., & Mokhlis, A. (2016a). A framework for the integration of Green and Lean Six Sigma for superior sustainability performance. *International Journal of Production Research, 55*(15), 4481–4515. https://doi.org/10.1080/00207543.2016.1266406

Cherrafi, A., Elfezazi, S., Chiarini, A., Mokhlis, A., & Benhida, K. (2016b). The integration of lean manufacturing, Six Sigma and sustainability: A literature review and future research directions for developing a specific model. *Journal of Cleaner Production, 139,* 828–846. https://doi.org/10.1016/j.jclepro.2016.08.101

Cherrafi, A., Elfezazi, S., Govindan, K., Garza-Reyes, J. A., Benhida, K., & Mokhlis, A. (2017). A framework for the integration of Green and Lean Six Sigma for superior sustainability performance. *International Journal of Production Research, 55*(15), 4481–4515. https://doi.org/10.1080/00207543.2016.1266406

Crosby, B., & Badurdeen, F. (2022). Integrating lean and sustainable manufacturing principles for sustainable total productive maintenance (Sus-TPM). *Smart and Sustainable Manufacturing Systems, 6*(1), 68–84.

Cucurachi, S., & Suh, S. (2017). Cause-effect analysis for sustainable development policy. *Environmental Reviews, 25*(3), 358–379.

da Silva, G. J., & Borges, A. C. (2025). Statistical Process Control in the Environmental Monitoring of Water Quality and Wastewaters: A Review. *Water, 17*(9), 1281.

Dahab, A., Younes, M. A., & Backar, S. (2024). Deployment of Lean Six Sigma framework with environmental considerations in SMEs: a case study. *International Journal of Productivity and Quality Management, 41*(3), 389–417.

de Freitas, J. G., & Costa, H. G. (2017). Impacts of Lean Six Sigma over organizational sustainability. *International Journal of Lean Six Sigma, 8*(1), 89–108. https://doi.org/10.1108/ijlss-10-2015-0039

Deepak, S., & Dharmendra, S. (2025). Green Lean Six Sigma Barriers and Efficacy in Ceramic Manufacturing Industries: A PLS-SEM Study. *Evergreen*, *12*(1), 176–190. https://doi.org/10.5109/7342448

Duarte, S., & Cruz-Machado, V. (2014). Investigating lean and green supply chain linkages through a balanced scorecard framework. *International Journal of Management Science and Engineering Management*, *10*(1), 20–29. https://doi.org/10.1080/17509653.2014.962111

Ebrahimi, A., Khakpour, R., & Saghiri, S. (2023). Sustainable setup stream mapping (3SM): a systematic approach to lean sustainable manufacturing. *Production Planning & Control*, *34*(4), 311–329.

Famiyeh, S., Adaku, E., Amoako-Gyampah, K., Asante-Darko, D., & Amoatey, C. T. (2018). Environmental management practices, operational competitiveness and environmental performance. *Journal of Manufacturing Technology Management*, *29*(3), 588–607. https://doi.org/10.1108/jmtm-06-2017-0124

Fargani, H., Cheung, W. M., & Hasan, R. (2016). An Empirical Analysis of the Factors That Support the Drivers of Sustainable Manufacturing. *Procedia CIRP*, *56*, 491–495. https://doi.org/10.1016/j.procir.2016.10.096

Farrukh, A. (2025). A comparative exploratory study of the critical success factors of a green-lean-six sigma strategy. *Business process Management journal*, 1–35. https://doi.org/10.1108/BPMJ-09-2024-0862

Farrukh, A., Mathrani, S., & Sajjad, A. (2021a). A natural resource and institutional theory-based view of green-lean-six sigma drivers for environmental management. *Business Strategy and the Environment*, *n/a*(n/a). https://doi.org/10.1002/bse.2936

Farrukh, A., Mathrani, S., & Sajjad, A. (2021b). A natural resource and institutional theory-based view of green-lean-six sigma drivers for environmental management. *Business Strategy and the Environment*, *31*(3), 1074–1090. https://doi.org/10.1002/bse.2936

Farrukh, A., Mathrani, S., & Sajjad, A. (2022a). Managerial perspectives on green-lean-six sigma adoption in the flexible packaging industry: empirical evidence from an emerging economy. *Journal of Manufacturing Technology Management*, *ahead-of-print*(ahead-of-print). https://doi.org/10.1108/JMTM-02-2022-0080

Farrukh, A., Mathrani, S., & Sajjad, A. (2022b). A Systematic Literature Review on Environmental Sustainability Issues of Flexible Packaging: Potential Pathways for Academic Research and Managerial Practice. *Sustainability*, *14*(8). https://doi.org/10.3390/su14084737

Farrukh, A., Mathrani, S., & Sajjad, A. (2023a). An exploratory study of green-lean-six sigma motivators for environmental sustainability: Managerial insights from a developed and developing economy. *Business Strategy and the Environment*. https://doi.org/10.1002/bse.3412

Farrukh, A., Mathrani, S., & Sajjad, A. (2023b). Green-lean-six sigma practices and supporting factors for transitioning towards circular economy: A natural resource and intellectual capital-based view. *Resources Policy*, *84*. https://doi.org/10.1016/j.resourpol.2023.103789

Farrukh, A., Mathrani, S., & Taskin, N. (2019, 9–11th December 2019). *Success Factors of a Combined Green, Lean, and Six Sigma Strategy for Environmental Performance* 5th i-COSTE 2019, the International Conference on Sustainable Technology and Engineering, CQUniversity, Melbourne, Australia.

Farrukh, A., Mathrani, S., & Taskin, N. (2020). Investigating the Theoretical Constructs of a Green Lean Six Sigma Approach towards Environmental Sustainability: A Systematic Literature Review and Future Directions. *Sustainability*, *12*(19). https://doi.org/10.3390/su12198247

Farrukh, A., & Sajjad, A. (2023). Investigating sustainability tensions and resolution strategies in the plastic food packaging industry—A paradox theory approach. *Business Strategy and the Environment*. https://doi.org/10.1002/bse.3637

Fayyaz, A., Liu, C., Xu, Y., Farooq, A., & Ahmed, S. (2025). Linkages between lean six sigma, green manufacturing, circular economy and operational performance in manufacturing organizations of a developing country. *International Journal of Lean Six Sigma*.

Fayyaz, A., Liu, C., Xu, Y., Khan, F., & Ahmed, S. (2024). Untangling the cumulative impact of big data analytics, green lean six sigma and sustainable supply chain management on the economic performance of manufacturing organisations. *Production Planning & Control, 36*(8), 1137–1154. https://doi.org/10.1080/09537287.2024.2348517

Fernando, A., Gunasekara, C., Siriwardana, C., Law, D., Zhang, G. K., & Gamage, J. (2025). Evaluating the Environmental Performance of Blended Cement-Based Concrete: A Comparative Life-Cycle Analysis to Derive an Environmental Score. *Journal of Materials in Civil Engineering, 37*(5), 04025093.

Fernando, P. U. A. I., Kennedy, A. J., Pokrzywinski, K., Jernberg, J., Thornell, T., George, G.,...Coyne, K. J. (2024). Development of alginate beads for precise environmental release applications: A design of experiment based approach and analysis. *Journal of Environmental Management, 351*, 119872.

Gaikwad, L., & Sunnapwar, V. (2020). An integrated Lean, Green and Six Sigma strategies. *The TQM Journal, 32*(2), 201–225. https://doi.org/10.1108/tqm-08-2018-0114

Gandhi, J., Thanki, S., & Thakkar, J. J. (2021). An investigation and implementation framework of Lean Green and Six Sigma (LG&SS) strategies for the manufacturing industry in India. *The TQM Journal, ahead-of-print*(ahead-of-print). https://doi.org/10.1108/tqm-12-2020-0289

Gandhi, N. S., Thanki, S. J., & Thakkar, J. J. (2018). Ranking of drivers for integrated lean-green manufacturing for Indian manufacturing SMEs. *Journal of Cleaner Production, 171*, 675–689. https://doi.org/10.1016/j.jclepro.2017.10.041

Garza-Reyes, J. A., Kumar, V., Chaikittisilp, S., & Tan, K. H. (2018a). The effect of lean methods and tools on the environmental performance of manufacturing organisations. *International Journal of Production Economics, 200*, 170–180. https://doi.org/10.1016/j.ijpe.2018.03.030

Garza-Reyes, J. A., Torres Romero, J., Govindan, K., Cherrafi, A., & Ramanathan, U. (2018b). A PDCA-based approach to Environmental Value Stream Mapping (E-VSM). *Journal of Cleaner Production, 180*, 335–348. https://doi.org/10.1016/j.jclepro.2018.01.121

Garza-Reyes, J. A., Yu, M., Kumar, V., & Upadhyay, A. (2018c). Total quality environmental management: adoption status in the Chinese manufacturing sector. *The TQM Journal, 30*(1), 2–19. https://doi.org/10.1108/tqm-05-2017-0052

Ghadimi, P., O'Neill, S., Wang, C., & Sutherland, J. W. (2020). Analysis of enablers on the successful implementation of green manufacturing for Irish SMEs. *Journal of Manufacturing Technology Management, 32*(1), 85–109. https://doi.org/10.1108/jmtm-10-2019-0382

Ghasemibojd, F., Franchetti, M. J., & George, B. (2025). Green lean six sigma for sustainable development: a systematic review of evolution, challenges, and future pathways. *Clean Technologies and Environmental Policy*. https://doi.org/10.1007/s10098-025-03281-y

Gholami, H., Jamil, N., Mat Saman, M. Z., Streimikiene, D., Sharif, S., & Zakuan, N. (2020). The application of Green Lean Six Sigma. *Business Strategy and the Environment.*

Ghosh, S., Mandal, M. C., & Ray, A. (2023). A PDCA based approach to evaluate green supply chain management performance under fuzzy environment. *International Journal of Management Science and Engineering Management, 18*(1), 1–15.

Ghosh, S., Mandal, M. C., & Ray, A. (2025). A PDCA based approach to evaluate green supply chain management performance under fuzzy environment. In *Urban Governance* (pp. 144–158). Routledge.

Hallioui, A., Herrou, B., Katina, P. F., Santos, R. S., Egbue, O., Jasiulewicz-Kaczmarek, M.,...Marques, P. C. (2023). A Review of Sustainable Total Productive Maintenance (STPM). *Sustainability, 15*(16), 12362.

Hariyani, D., Hariyani, P., Mishra, S., & Sharma, M. K. (2025). Drivers for the adoption of integrated sustainable green lean six sigma agile service system (ISGLSASS) in the service organizations. *Sustainable Futures, 9*. https://doi.org/10.1016/j.sftr.2025.100571

Hariyani, D., & Mishra, S. (2022a). Barriers to the adoption of integrated sustainable-green-lean-six sigma-agile manufacturing system (ISGLSAMS): a literature review. *Benchmarking: An International Journal, 30*(9), 3590–3636. https://doi.org/10.1108/bij-10-2021-0585

Hariyani, D., & Mishra, S. (2022b). Drivers for the adoption of integrated sustainable green lean six sigma agile manufacturing system (ISGLSAMS) and research directions. *Cleaner engineering and technology, 7*, 100449.

Hariyani, D., Mishra, S., Sharma, M. K., & Hariyani, P. (2023). A study of the barriers to the adoption of integrated sustainable-green-lean-six sigma-agile manufacturing system (ISGLSAMS) in Indian manufacturing organizations. *Cleaner Waste Systems, 5*. https://doi.org/10.1016/j.clwas.2023.100098

Hassan, N. A. E.-M., Metwally, F. A., & Abdul-Samee, A. N. (2025). Using Pareto Diagram to Improve Textile Product Quality: A Case Study. *International Design Journal, 15*(4), 345-350.

Horodytska, O., Kiritsis, D., & Fullana, A. (2020). Upcycling of printed plastic films: LCA analysis and effects on the circular economy. *Journal of Cleaner Production, 268*. https://doi.org/10.1016/j.jclepro.2020.122138

Horsthofer-Rauch, J. (2025). Sustainability-Focused Value Stream Mapping Using Process Mining.

Hudnurkar, M., Ambekar, S., & Bhattacharya, S. (2019). Empirical analysis of Six Sigma project capability deficiency and its impact on project success. *The TQM Journal, 31*(3), 340–358. https://doi.org/10.1108/tqm-06-2018-0078

Hussain, K., He, Z., Ahmad, N., Iqbal, M., & Nazneen, S. (2022). Mapping Green, Lean, Six Sigma enablers through the lens of a construction sector: an emerging economy's perspective. *Journal of Environmental Planning and Management*, 1–34. https://doi.org/10.1080/09640568.2021.2006155

Hussain, K., He, Z., Ahmad, N., Iqbal, M., & Saeed, M. Z. (2022). Establishing a Green, Lean and Six Sigma implementation model for sustainable construction industry: an analysis of driving forces through ISM-MICMAC approach. *Environ Sci Pollut Res Int*. https://doi.org/10.1007/s11356-022-24039-9

Hussain, K., He, Z., Ahmad, N., Iqbal, M., & Taskheer mumtaz, S. M. (2019). Green, lean, Six Sigma barriers at a glance: A case from the construction sector of Pakistan. *Building and Environment, 161*. https://doi.org/10.1016/j.buildenv.2019.106225

Iliná, A., Ramos-González, R., Arredondo-Valdés, R., Barrera-Martínez, C., Laredo-Alcalá, E., Albarracin, P. M.,...Martínez-Hernández, J. L. (2023). Food Waste Management Method Through 3R Concept. In *Food Waste Conversion* (pp. 43–52). Springer.

Iqbal, T., Huq, F., & Bhutta, M. K. S. (2018). Agile manufacturing relationship building with TQM, JIT, and firm performance: An exploratory study in apparel export industry of Pakistan. *International Journal of Production Economics, 203*, 24–37. https://doi.org/10.1016/j.ijpe.2018.05.033

Ishak, A., Mohamad, E., & Rahman, M. A. A. (2024). Confirmatory factor analysis for the implementation of Green Lean Six Sigma (GLSS) in Malaysia's wastewater treatment industry. *Water Policy*. https://doi.org/10.2166/wp.2024.095

Jamil, N., Gholami, H., Saman, M. Z. M., Streimikiene, D., Sharif, S., & Zakuan, N. (2020). DMAIC-based approach to sustainable value stream mapping: towards a sustainable manufacturing system. *Economic Research-Ekonomska Istraživanja, 33*(1), 331–360. https://doi.org/10.1080/1331677x.2020.1715236

Jin, H., Qu, P., Zhou, Y., Qu, Q., & Li, H. (2024). Green Lean Six Sigma practice in the industrial and service sectors: from the systematic literature review to an integration diagram. *Production Planning & Control*, 1–26. https://doi.org/10.1080/09537287.2024.2369779

Jonda, E., Karkoszka, T., & Jonda, K. (2023). Recycling materials database as a Green Kaizen for sustainable development in the automotive industry. *Journal of Achievements in Materials and Manufacturing Engineering, 120*(1).

Jovanović, B., Filipović, J., & Bakić, V. (2017). Energy management system implementation in Serbian manufacturing—Plan-Do-Check-Act cycle approach. *Journal of Cleaner Production, 162*, 1144–1156. https://doi.org/10.1016/j.jclepro.2017.06.140

Kaswan, M. S., & Rathi, R. (2019). Analysis and modeling the enablers of Green Lean Six Sigma implementation using Interpretive Structural Modeling. *Journal of Cleaner Production, 231*, 1182–1191. https://doi.org/10.1016/j.jclepro.2019.05.253

Kaswan, M. S., & Rathi, R. (2020a). Green Lean Six Sigma for sustainable development: Integration and framework. *Environmental Impact Assessment Review, 83*. https://doi.org/10.1016/j.eiar.2020.106396

Kaswan, M. S., & Rathi, R. (2020b). Investigating the enablers associated with implementation of Green Lean Six Sigma in manufacturing sector using Best Worst Method. *Clean Technologies and Environmental Policy*, 1–12.

Kaswan, M. S., Rathi, R., Antony, J., Cross, J., Garza-Reyes, J. A., Singh, M.,...Sony, M. (2024). Integrated Green Lean Six Sigma-Industry 4.0 approach to combat COVID-19: from literature review to framework development. *International Journal of Lean Six Sigma, 15*(1), 50–79. https://doi.org/10.1108/IJLSS-11-2022-0227

Kaswan, M. S., Rathi, R., Reyes, J. A. G., & Antony, J. (2021). Exploration and Investigation of Green Lean Six Sigma Adoption Barriers for Manufacturing Sustainability. *IEEE Transactions on Engineering Management*, 1–15. https://doi.org/10.1109/tem.2021.3108171

Kattumuri, R. (2018). Sustaining natural resources in a changing environment: evidence, policy and impact. In: Taylor & Francis.

Kennelly, C., Berners-Lee, M., & Hewitt, C. N. (2019). Hybrid life-cycle assessment for robust, best-practice carbon accounting. *Journal of Cleaner Production, 208*, 35–43. https://doi.org/10.1016/j.jclepro.2018.09.231

Khamidov, I., Costa, F., & Staudacher, A. P. (2023). How Lifecycle Assessment is Interrelated with Environmental or Sustainable Value Stream Mapping.

Klein, L. L., Moyano-Fuentes, J., Vieira, K. M., & Marçal, D. R. (2024). An exploratory study of the relationships between Lean practices and team performance in higher education. *International Journal of Lean Six Sigma, 15*(2), 372–399. https://doi.org/10.1108/IJLSS-08-2022-0168

Kumar, M., & Rodrigues, V. S. (2020). Synergetic effect of lean and green on innovation: A resource-based perspective. *International Journal of Production Economics, 219*, 469–479. https://doi.org/10.1016/j.ijpe.2018.04.007

Kumar, S., Luthra, S., Govindan, K., Kumar, N., & Haleem, A. (2016). Barriers in green lean six sigma product development process: an ISM approach. *Production Planning & Control*, 1–17. https://doi.org/10.1080/09537287.2016.1165307

Kurnia, H., Jaqin, C., & Purba, H. H. (2022). Quality improvement with PDCA approach and design of experiment method in single socks industry in Indonesia.

Lara, A. C., Menegon, E. M. P., Sehnem, S., & Kuzma, E. (2022). Relationship between just in time, lean manufacturing, and performance practices: a meta-analysis. *Gestão & Produção, 29*, e9021.

Larsson, D., Ratnayake, R. M. C., & Samarakoon, S. M. K. (2025). Enhancing Performance in Engineering-To-Order Projects: Integrating Digital Value Stream Mapping and Green Lean Practices. *Engineering Management Journal*, 1–18.

Lee, V.-H., Ooi, K.-B., Chong, A. Y.-L., & Lin, B. (2015). A structural analysis of greening the supplier, environmental performance and competitive advantage. *Production Planning & Control, 26*(2), 116–130.

Letchumanan, L. T., Gholami, H., Yusof, N. M., Ngadiman, N. H. A. B., Salameh, A. A., Štreimikienė, D., & Cavallaro, F. (2022). Analyzing the Factors Enabling Green Lean Six Sigma Implementation in the Industry 4.0 Era. *Sustainability, 14*(6). https://doi.org/10.3390/su14063450

Lino Moreno, S. E., Navarro Ayola, B. L., Salas, R., & Nallusamy, S. (2025). Enhancement of Operational Efficiency in a Plastic Manufacturing Industry Through TPM, SMED, and Machine Learning—Case Study. *Sustainability, 17*(16), 7445.

Liu, J., Ma, Y., Appolloni, A., & Cheng, W. (2021). How external stakeholders drive the green public procurement practice? An organizational learning perspective. *Journal of Public Procurement, 21*(2), 138–166. https://doi.org/10.1108/jopp-04-2020-0035

Lizarelli, F. L., Antony, J., Suarez, M., Carneiro, F., Sony, M., Chakraborty, A.,...Chan, F. T. S. (2025). Analysis of the impact of Kaizen practices on ESG performance and the mediating role of digital systems. *Business process Management journal, 31*(8), 148–175.

Lizarelli, F. L., Chakraborty, A., Antony, J., Furterer, S., Maalouf, M., & Carneiro, M. B. (2024). An empirical study on Lean and its impact on sustainability in services. *International Journal of Quality & Reliability Management, 41*(3), 805–825.

Masri, H. A., & Jaaron, A. A. M. (2017). Assessing green human resources management practices in Palestinian manufacturing context: An empirical study. *Journal of Cleaner Production, 143*, 474–489. https://doi.org/10.1016/j.jclepro.2016.12.087

McDermott, O., Moloney, C., Noonan, J., & Rosa, A. (2024). Green Lean Six Sigma in the food industry: a systematic literature review. *British Food Journal, 126*(13), 455–469. https://doi.org/10.1108/bfj-01-2024-0100

McGlinchey, Y., Iqbal, J., Trubetskaya, A., Thenarasu, M., & McDermott, O. (2025). Design for green lean six sigma to improve sustainability in the pharmaceutical industry—A case study. *Production & Manufacturing Research, 13*(1). https://doi.org/10.1080/21693277.2025.2461010

Mishra, M. N. (2018). Identify critical success factors to implement integrated green and Lean Six Sigma. *International Journal of Lean Six Sigma.*

Mittal, V. K., & Sangwan, K. S. (2015). Ranking of Drivers for Green Manufacturing Implementation Using Fuzzy Technique for Order of Preference by Similarity to Ideal Solution Method. *Journal of Multi-Criteria Decision Analysis, 22*(1–2), 119–130. https://doi.org/10.1002/mcda.1527

Mohammad, A., Hamja, A., & Hasle, P. (2024). Reduction of changeover time through SMED with RACI integration in garment factories. *International Journal of Lean Six Sigma, 15*(2), 201–219. https://doi.org/10.1108/IJLSS-10-2021-0176

Mohammed, M., Shafiq, N., Elmansoury, A., Al-Mekhlafi, A.-B. A., Rached, E. F., Zawawi, N. A.,…Ibrahim, M. B. (2021). Modeling of 3R (reduce, reuse and recycle) for sustainable construction waste reduction: A partial least squares structural equation modeling (PLS-SEM). *Sustainability, 13*(19), 10660.

Mohan, J., Kaswan, M. S., & Rathi, R. (2023). Identification and investigation into the barriers to Green Lean Six Sigma implementation: a micro small and medium enterprises perspective. *International Journal on Interactive Design and Manufacturing (IJIDeM).* https://doi.org/10.1007/s12008-023-01551-x

Mohan, J., Kaswan, M. S., & Rathi, R. (2024a). An analysis of green lean six sigma deployment in MSMEs: a systematic literature review and conceptual implementation framework. *The TQM Journal.*

Mohan, J., Kaswan, M. S., & Rathi, R. (2024b). An analysis of green lean six sigma deployment in MSMEs: a systematic literature review and conceptual implementation framework. *The TQM Journal.* https://doi.org/10.1108/tqm-06-2023-0197

Mohan, J., Kaswan, M. S., & Rathi, R. (2025a). An analysis of green lean six sigma deployment in MSMEs: a systematic literature review and conceptual implementation framework. *The TQM Journal, 37*(3), 747–777. https://doi.org/10.1108/TQM-06-2023-0197

Mohan, J., Kaswan, M. S., & Rathi, R. (2025b). Assessing enablers of Green Lean Six Sigma: lessons from Indian micro, small, and medium enterprises [Article]. *International Journal of Six Sigma and Competitive Advantage, 15*(3), 333–350. https://doi.org/10.1504/IJSSCA.2025.145632

Mollenkopf, D., Halldórsson, Á., Stolze, H., Tate, W. L., & Ueltschy, M. (2010). Green, lean, and global supply chains. *International Journal of Physical Distribution & Logistics Management, 40*(1/2), 14–41. https://doi.org/10.1108/09600031011018028

Morales-García, A. S., Díaz-Reza, J. R., & García-Alcaraz, J. L. (2021). Effect of TPM and OEE on the social performance of companies. In *New perspectives on enterprise decision-making applying artificial intelligence techniques* (pp. 119–141). Springer.

Nababan, E., Siahaan, N., Bangun, P., & Rosmaini, E. (2018). Environmental Performance Measurement of the Simple Urban Housing in Martubung Medan.

Nguyen, D. T., & Sharmak, W. (2022). Hybrid Simulation for Value Stream Mapping to Improve the Environmental Performance of the Construction Phase.

Nkansah, C. (2025). Leveraging Six Sigma for Continuous Improvement in Petroleum and Chemical Processing. *International Journal of Advance Industrial Engineering, 13*(01).

Ong, T. S., Lee, A. S., Latif, B., Sroufe, R., Sharif, A., & Heng Teh, B. (2023). Enabling green shared vision: linking environmental strategic focus and environmental performance through ISO 14001 and technological capabilities. *Environmental Science and Pollution Research*, *30*(11), 31711–31726.

Orji, I. J. (2019). Examining barriers to organizational change for sustainability and drivers of sustainable performance in the metal manufacturing industry. *Resour. Conserv. Recycl.*, *140*, 102–114. https://doi.org/10.1016/j.resconrec.2018.08.005

Paliska, G., Pavletic, D., & Sokovic, M. (2007). Quality tools–systematic use in process industry. *Journal of Achievements in Materials and Manufacturing Engineering*, *25*(1), 79–82.

Panwar, A., Nepal, B., Jain, R., Rathore, A. P. S., & Lyons, A. (2017). Understanding the linkages between lean practices and performance improvements in Indian process industries. *Industrial Management & Data Systems*, *117*(2), 346–364. https://doi.org/10.1108/imds-01-2016-0035

Parmar, P. S., & Desai, T. N. (2020a). Evaluating Sustainable Lean Six Sigma enablers using fuzzy DEMATEL: A case of an Indian manufacturing organization. *Journal of Cleaner Production*, *265*. https://doi.org/10.1016/j.jclepro.2020.121802

Parmar, P. S., & Desai, T. N. (2020b). Ranking the solutions of Sustainable Lean Six Sigma implementation in Indian manufacturing organization to overcome its barriers. *International Journal of Sustainable Engineering*, *14*(3), 304–317. https://doi.org/10.1080/19397038.2020.1813834

Perez-Burgoin, M., Baez-Lopez, Y., Limon-Romero, J., Tlapa, D., & García-Alcaraz, J. L. (2024). Enablers for green lean six sigma adoption in the manufacturing industry. *Journal of Manufacturing Technology Management*, *35*(6), 1199–1225. https://doi.org/10.1108/JMTM-09-2023-0396

Rahardjo, B., Wang, F.-K., Yeh, R.-H., & Chen, Y.-P. (2023). Lean manufacturing in industry 4.0: a smart and sustainable manufacturing system. *Machines*, *11*(1), 72.

Rahardjo, B., Winnyarto, S. T., & Tjendra, V. (2025). Integration of green lean six sigma and TRIZ methodology for sustainable green manufacturing: a case study in plywood industry. *International Journal of Lean Six Sigma*. https://doi.org/10.1108/ijlss-09-2024-0201

Rashid, A., Rasheed, R., & Amirah, N. A. (2025). Synergizing TQM, JIT, and Green Supply Chain Practices: Strategic Insights for Enhanced Environmental Performance. *Logistics*, *9*(1), 18.

Rathi, R., Kaswan, M. S., Antony, J., Cross, J., Garza-Reyes, J. A., & Furterer, S. L. (2023). Success factors for the adoption of green lean six sigma in healthcare facility: an ISM-MICMAC study. *International Journal of Lean Six Sigma*, *14*(4), 864–897. https://doi.org/10.1108/IJLSS-02-2022-0042

Rathi, R., Kaswan, M. S., Garza-Reyes, J. A., Antony, J., & Cross, J. (2022). Green Lean Six Sigma for improving manufacturing sustainability: Framework development and validation. *Journal of Cleaner Production*, *345*. https://doi.org/10.1016/j.jclepro.2022.131130

Rathi, R., Vakharia, A., & Kaswan, M. S. (2021). Grey relational analysis of Green Lean Six Sigma critical success factors for improved organisational performance. *International Journal of Six Sigma and Competitive Advantage*, *13*(1–3), 55–75.

Ren, J., Manzardo, A., Toniolo, S., & Scipioni, A. (2013). Sustainability of hydrogen supply chain. Part I: Identification of critical criteria and cause–effect analysis for enhancing the sustainability using DEMATEL. *International journal of hydrogen energy*, *38*(33), 14159–14171.

Riaz, H., Saeed, A., Liedong, T. A., & Rajwani, T. (2022). Environmental management, nonmarket strategy, and firm performance in emerging markets: The case of ISO 14001. *Business Ethics, the Environment & Responsibility*, *31*(1), 139–163.

Rodriguez Delgadillo, R., Medini, K., & Wuest, T. (2022). A DMAIC framework to improve quality and sustainability in additive manufacturing—A case study. *Sustainability*, *14*(1), 581.

Sajjad, A., Eweje, G., & Tappin, D. (2019). Managerial perspectives on drivers for and barriers to sustainable supply chain management implementation: Evidence from New Zealand. *Business Strategy and the Environment*, *29*(2), 592–604. https://doi.org/10.1002/bse.2389

Samadhiya, A., & Agrawal, R. (2024). Total productive maintenance and sustainability performance: resource-based view perspective. *Benchmarking: An International Journal*, *31*(7), 2177–2196.

Samadhiya, A., Agrawal, R., Kumar, A., & Garza-Reyes, J. A. (2023). Blockchain technology and circular economy in the environment of total productive maintenance: a natural resource-based view perspective. *Journal of Manufacturing Technology Management*, *34*(2), 293–314.

Santoso, I., Rau, H., Choirun, A. u., Aprilianto, H. C., Lestari, K., A'yuniah, S., & Kusumaningtyas, O. W. (2024). Application of QFD in sustainable new product development in the agro-industrial sector: a systematic literature review. *International Journal of Sustainable Engineering*, *17*(1), 926–944.

Seth, D., & Rehman, M. A. A. (2022). Critical success factors-based strategy to facilitate green manufacturing for responsible business: An application experience in Indian context. *Business Strategy and the Environment*, *31*(7), 2786–2806. https://doi.org/10.1002/bse.3047

Seth, D., Rehman, M. A. A., & Shrivastava, R. L. (2018). Green manufacturing drivers and their relationships for small and medium(SME) and large industries. *Journal of Cleaner Production*, *198*, 1381–1405. https://doi.org/10.1016/j.jclepro.2018.07.106

Setiawan, N., Salleh, M. R., Ariff, H. A., Rahman, M. A. A., Mohamad, E., Sulaiman, M. A.,…Ito, T. (2021). A proposal of performance measurement and management model for 5S sustainability in manufacturing SMEs: A Review. *Journal of Advanced Mechanical Design, Systems, and Manufacturing*, *15*(2), JAMDSM0017–JAMDSM0017.

Sharma, D., & Singh, D. (2025). Green Lean Six Sigma Barriers and Efficacy in Ceramic Manufacturing Industries: A PLS-SEM Study.

Sharma, V., Gidwani, B. D., Sharma, V., & Meena, M. L. (2022). Causal relationship among critical factors for cellular manufacturing system using DEMATEL approach. *International Journal of Business and Systems Research*, *16*(3), 277–301.

Shokri, A., Antony, J., & Garza-Reyes, J. A. (2022). A new way of environmentally sustainable manufacturing with assessing transformation through the green deployment of Lean Six Sigma projects. *Journal of Cleaner Production*, *351*. https://doi.org/10.1016/j.jclepro.2022.131510

Shokri, A., & Li, G. (2020). Green implementation of Lean Six Sigma projects in the manufacturing sector. *International Journal of Lean Six Sigma*, *11*(4), 711–729.

Shrivastava, A., & Mishra, R. P. (2025). Deploying lean six sigma and industry 4.0 framework in an auto motive manufacturing organization for establishing circular economy. *OPSEARCH*, 1–32.

Siegel, R., Antony, J., Garza-Reyes, J. A., Cherrafi, A., & Lameijer, B. (2019). Integrated green lean approach and sustainability for SMEs: From literature review to a conceptual framework. *Journal of Cleaner Production*, *240*. https://doi.org/10.1016/j.jclepro.2019.118205

Silva, B. B., Mergulhão, R. C., Favoretto, C., & Mendes, G. H. S. (2019). Critical success factors of Six Sigma implementations in companies in Brazil. *International Journal of Lean Six Sigma*, *10*(1), 143–160. https://doi.org/10.1108/ijlss-01-2018-0008

Singh, M., Rathi, R., & Garza-Reyes, J. A. (2021). Analysis and prioritization of Lean Six Sigma enablers with environmental facets using best worst method: A case of Indian MSMEs. *Journal of Cleaner Production*, *279*. https://doi.org/10.1016/j.jclepro.2020.123592

Singh Sangwan, K., Bhamu, J., & Mehta, D. (2014). Development of lean manufacturing implementation drivers for Indian ceramic industry. *International Journal of Productivity and Performance Management*, *63*(5), 569–587. https://doi.org/10.1108/ijppm-06-2013-0105

Singhal, V., Maiyar, L. M., & Roy, I. (2025). Environmental sustainability consideration with just-in-time practices in industry 4.0 era–A state of the art. *Operations Management Research*, *18*(2), 437–460.

Siwiec, D., & Pacana, A. (2025). Sustainable Prototyping: Linking Quality and Environmental Impact via QFD and LCA. *Sustainability*, *17*(13), 5818.

Sodhi, H. S., Singh, D., & Singh, B. J. (2020). A conceptual examination of Lean, Six Sigma and Lean Six Sigma models for managing waste in manufacturing SMEs. *World Journal of Science, Technology and Sustainable Development*, *17*(1), 20–32. https://doi.org/10.1108/wjstsd-10-2019-0073

Sony, M., Antony, J., & Naik, S. (2020). How do organizations implement an effective LSS initiative? A qualitative study. *Benchmarking: An International Journal*.

Soti, A., Shankar, R., & Kaushal, O. P. (2010). Modeling the enablers of Six Sigma using interpreting structural modeling. *Journal of Modelling in Management, 5*(2), 124–141. https://doi.org/10.1108/17465661011060989

Sreedharan, R., Sunder M, V., & R, R. (2018). Critical success factors of TQM, Six Sigma, Lean and Lean Six Sigma. *Benchmarking: An International Journal, 25*(9), 3479–3504. https://doi.org/10.1108/bij-08-2017-0223

Subramanian, N., & Abdulrahman, M. (2017). An examination of drivers and barriers to reducing carbon emissions in China's manufacturing sector. *The International Journal of Logistics Management, 28*(4), 1168–1195.

Swarnakar, V., & Khalfan, M. (2025). Unveiling the impact of operational excellence practices on corporate performance: an exploratory research. *The TQM Journal.*

Swarnakar, V., Tiwari, A. K., & Singh, A. R. (2020). Evaluating critical failure factors for implementing sustainable lean six sigma framework in manufacturing organization. *International Journal of Lean Six Sigma, 11*(6), 1083–1118. https://doi.org/10.1108/ijlss-05-2019-0050

Tan, D., Wu, Y., Zhang, Z., Jiao, Y., Zeng, L., & Meng, Y. (2023). Assessing the life cycle sustainability of solar energy production systems: A toolkit review in the context of ensuring environmental performance improvements. *Sustainability, 15*(15), 11724.

Tasdemir, C., Gazo, R., & Quesada, H. J. (2019). Sustainability benchmarking tool (SBT): theoretical and conceptual model proposition of a composite framework. *Environment, Development and Sustainability.* https://doi.org/10.1007/s10668-019-00512-3

Thanki, Govindan, K., & Thakkar, J. (2016). An investigation on lean-green implementation practices in Indian SMEs using analytical hierarchy process (AHP) approach. *Journal of Cleaner Production, 135*, 284–298. https://doi.org/10.1016/j.jclepro.2016.06.105

Thomas, D., & Khanduja, D. (2021). ISM–ANP hybrid approach to prioritize the barriers in green lean Six Sigma implementation in construction sector. *International Journal of Lean Six Sigma, 13*(2), 502–520. https://doi.org/10.1108/ijlss-09-2020-0140

Tibebu, M., & Henry, M. (2025). Green Value Stream Mapping for Realizing Sustainability in Urban Road Construction using the DMAIC Approach.

Tjahjono, B., Ball, P., Vitanov, V. I., Scorzafave, C., Nogueira, J., Calleja, J., . . . Yadav, A. (2010). Six Sigma: a literature review. *International Journal of Lean Six Sigma, 1*(3), 216–233. https://doi.org/10.1108/20401461011075017

Tlapa, D., Limon, J., García-Alcaraz, J. L., Baez, Y., & Sánchez, C. (2016). Six Sigma enablers in Mexican manufacturing companies: a proposed model. *Industrial Management & Data Systems, 116*(5), 926–959. https://doi.org/10.1108/imds-06-2015-0265

Trianni, A., Cagno, E., & Farné, S. (2016). Barriers, drivers and decision-making process for industrial energy efficiency: A broad study among manufacturing small and medium-sized enterprises. *Applied Energy, 162*, 1537–1551.

Tsarouhas, P., & Sidiropoulou, N. (2024). Application of Six Sigma methodology using DMAIC approach for a packaging olives production system: a case study. *International Journal of Lean Six Sigma, 15*(2), 247–273. https://doi.org/10.1108/IJLSS-06-2022-0140

Wałach, D. (2020). Analysis of factors affecting the environmental impact of concrete structures. *Sustainability, 13*(1), 204.

Walker, H., Di Sisto, L., & McBain, D. (2008). Drivers and barriers to environmental supply chain management practices: Lessons from the public and private sectors. *Journal of Purchasing and Supply Management, 14*(1), 69–85. https://doi.org/10.1016/j.pursup.2008.01.007

Walter, O. M. F. C., Paladini, E. P., Henning, E., & Kalbusch, A. (2021). Recent developments in sustainable lean six sigma frameworks: literature review and directions. *Production Planning & Control*, 1–19. https://doi.org/10.1080/09537287.2021.1967497

Wang, C.-N., Vo, T. T. B. C., Le, D.-S., & Chung, Y.-C. (2025). Optimizing Production Line Efficiency via DMAIC, VSM and FMEA: A Case Study of an Electronics Manufacturing Services Company. *Engineering Management Journal*, 1–19.

Wessels, J., & Ross, C. (2024). Evaluating the contribution of ISO 14001: 2015 towards enhancement of environmental performance: A mining case study.

Yadav, G., Luthra, S., Huisingh, D., Mangla, S. K., Narkhede, B. E., & Liu, Y. (2020). Development of a lean manufacturing framework to enhance its adoption within manufacturing companies in developing economies. *Journal of Cleaner Production, 245*. https://doi.org/10.1016/j.jclepro.2019.118726

Yadav, V., & Gahlot, P. (2022a). Green Lean Six Sigma sustainability-oriented framework for small and medium enterprises. *International Journal of Quality & Reliability Management*. https://doi.org/10.1108/ijqrm-08-2021-0297

Yadav, V., & Gahlot, P. (2022b). Green Lean Six Sigma sustainability-oriented framework for small and medium enterprises. *International Journal of Quality & Reliability Management, 39*(7), 1787–1807.

Yadav, V., Kaswan, M. S., Gahlot, P., Duhan, R. K., Garza-Reyes, J. A., Rathi, R.,...Yadav, G. (2023). Green Lean Six Sigma for sustainability improvement: a systematic review and future research agenda. *International Journal of Lean Six Sigma, 14*(4), 759–790. https://doi.org/10.1108/ijlss-06-2022-0132

Yadav, V., Kumar, V., Gahlot, P., Mittal, A., Kaswan, M. S., Garza-Reyes, J. A.,...Owad, A. A. (2023). Exploration and mitigation of green lean six sigma barriers: a higher education institutions perspective. *The TQM Journal*. https://doi.org/10.1108/tqm-03-2023-0069

Yalabik, B., & Fairchild, R. J. (2011). Customer, regulatory, and competitive pressure as drivers of environmental innovation. *International Journal of Production Economics, 131*(2), 519–527.

Yang, M., Chen, S., & Maresova, P. (2024). Environmental corporate social responsibility and stock price crash risk: The role of environmental performance and ISO 14001. *International Review of Economics & Finance, 96*, 103627.

Chapter 3
Theoretical Foundations of the Green-Lean-Six Sigma Approach

3.1 Introduction

This chapter discusses dominant organizational perspectives that explain GLSS adoption and implementation, with a specific focus on the drivers, barriers, tools, practices, and critical success factors for addressing organizations' environmental sustainability. From this context, theoretical lenses such as the natural resource-based view, intellectual capital-based view, institutional theory, and stakeholder theory are considered, aligning with the GLSS approach.

3.2 Theoretical Underpinnings of the Green-Lean-Six Sigma Approach

A suitable organizational theory helps in explicating "*why and how acts, events, and thoughts occur*" in relation to the chosen research phenomenon (Chang et al. 2017, p. 48). The extant literature revealed the use of organizational theories in the GLSS domain, such as the natural resource-based view (Afum et al. 2023; Mohamad et al. 2024), institutional theory (Abobakr et al. 2023; Braunscheidel et al. 2011; Caldera et al. 2019; Farrukh et al. 2023a), resource-based view (Adebanjo et al. 2016b), practice-based view (Al-Hakimi et al. 2025; Buhaya and Metwally 2024; Skalli et al. 2024; Tiwari et al. 2020), and an intellectual capital-based view (Farrukh 2025; Farrukh et al. 2023b). Since a holistic GLSS strategy encompasses various aspects such as motivators, tools, critical success factors, enablers, and barriers, an in-depth understanding of these issues requires multiple organizational theories. Using a single theory can lead to bias and fail to address the complex and dynamic nature of the GLSS concept (Farrukh 2022).

A. Farrukh and A. Sajjad, *A Manager's Guide to Green-Lean-Six Sigma Application*,
https://doi.org/10.1007/978-3-032-21111-8_3

3.2.1 Natural Resource-Based View of Green-Lean-Six Sigma

The resource-based view (RBV) emphasizes the significance of a firm's rare, valuable, inimitable, and distinct (i.e., non-substitutable) resources (Ferreira and Ferreira 2025; Khanra et al. 2022; Llach et al. 2025). Nurturing core competencies helps an organization to achieve a competitive advantage from a sustainability perspective (Barney 1991; Nayak et al. 2023). According to Hart (1995, p. 986), the RBV holds that "valuable, costly to copy firm resources and capabilities provide the key sources of sustainable competitive advantage". In addition, an RBV emphasizes that an organization can only sustain their competitive advantage if the resources supporting the firm's capabilities cannot be duplicated or pose "barriers to imitation" (Hart 1995, p. 987). The RBV also highlights that an organization's strategic resources must be of significant importance to a specific firm, and these should be rare (Beamish and Chakravarty 2021; Gueler and Schneider 2021). The RBV has been further extended by incorporating the environmental criteria and significance of conserving natural resources in the organizational context, which is broadly referred to as the natural resource-based view (NRBV) (Hart 1995).

Hart (1995, p. 986) argues that "historically, management theory has ignored the constraints imposed by the biophysical (natural) environment". The NRBV, however, addresses this issue by focusing on the importance of developing and conserving natural resource capabilities to achieve a sustainable competitive advantage. Under the NRBV, the key strategic capabilities of a firm include pollution prevention, product stewardship, and sustainable development (Cristina De Stefano et al. 2016; Hart 1995; Hart et al. 2010; Menguc and Ozanne 2005). The pollution prevention capability refers to reducing process waste and air emissions rather than cleaning up at the 'end of pipe'. It further reduces costs by streamlining processes and reducing input requirements, achieving environmental and economic sustainability (Gabriel et al. 2018; Lau and Wong 2024). Pollution prevention also improves an organization's productivity and efficiency by reducing waste, which further contributes to optimizing the utilization of resources such as raw materials and reducing the cost of material and waste disposal (Cristina De Stefano et al. 2016; Grama-Vigouroux et al. 2025; Smart 1992). The NRBV holds that by removing unnecessary, non-value-added steps and simplifying the production operations, firms can reduce their cycle time and lead time, achieving customer satisfaction (Stalk Jr and Hout 1990). Moreover, by reducing air emissions, pollution prevention capacities help organizations comply with environmental regulations, which leads to reducing environmental fines and liability costs (Foulon and Marsat 2023; Rooney 1993; Tian et al. 2023). Further, a pollution prevention capability can foster a firm's cash flow management and enhance overall profitability and productivity (Agyabeng-Mensah et al. 2021; Tan et al. 2022).

While pollution prevention focuses on reducing waste and emissions in a firm's production operations, a firm should also focus on addressing the environmental impacts across its entire value-chain activities from raw material access to disposal of the products, which the pollution prevention capability cannot address (Cobb and Daly 1989). Product stewardship refers to an organization's ability to address the external stakeholders' perspective by incorporating the 'voice of the natural environment' in its design and development stages (Fiksel 1993; Hart 1995). In particular, product stewardship capability focuses on the environmental sustainability issues during the product life cycle and promotes eco-friendly product design (Cristina De Stefano et al. 2016; Gabriel et al. 2018; Sharma et al. 2024). In this regard, organizations utilize environmental practices such as life cycle analysis to determine the environmental impacts of a product across different stages, such as from cradle to grave (Coppola et al. 2023; Keoleian 1993). It enables organizations to reduce raw material consumption, minimize hazardous materials, and promote sustainable and renewable resources, reducing life cycle costs (Alkaraan et al. 2024; Johnson-Hall and Hall 2022; Sehnem et al. 2022). Product stewardship also emphasizes the reuse, composting, remanufacturing, redesigning, and recycling of products (Kleiner 1991; Kühl et al. 2022; Mengistu et al. 2025; Yuan et al. 2023).

Meanwhile, sustainable development capability refers to implementing practices that can improve resource efficiency from a long-term perspective, reduce operational and environmental waste, and improve workplace safety (Alkaraan et al. 2024; Baumgartner and Rauter 2017; Liu 2025; Tiwari et al. 2020). It also emphasizes the "long-term vision to leverage an environmentally conscious strategy into the developing world that includes low-impact technology and products as the basis for market entry and development" (Hart 1995, p. 989). Additionally, using environmental technologies and eco-friendly and sustainable resources from a long-term perspective, an organization can achieve a competitive advantage through its sustainable development strategic capability (Grama-Vigouroux et al. 2025; Lichtenthaler 2022; Münch et al. 2022).

Regarding the GLSS strategy, an organization's internal motivations behind implementing this approach include cost reduction, waste reduction, improving productivity, achieving continuous improvement, and enhancing efficiency, which are aligned with the pollution prevention strategic capability (Abobakr et al. 2022; Hart et al. 2010). Similarly, a firm's motives for GLSS implementation stem from reducing life cycle cost, improving recyclability and reusability of raw materials, and achieving eco-design objectives associated with the product stewardship strategic capability (Afum et al. 2022; Farrukh et al. 2023a; Hamdoun 2020). Finally, the organization's interest in achieving long-term profitability and sustainability using the GLSS strategy are aligned with the strategic aspects of sustainable development from an NRBV perspective (Ahmad et al. 2025; Farrukh et al. 2021; Johnson-Hall and Hall 2022).

GLSS practices and tools relate to organizations' pollution prevention, product stewardship, and sustainable development strategic capabilities. For example, tools such as SMED and VSM improve the production flow by removing the non-value-added activities from the process, along with minimizing lead time, reducing waste, and minimizing air emissions (Ahmad et al. 2021; Chiarini 2014; Farrukh et al. 2022; Garza-Reyes et al. 2018). Similarly, practices such as eco-design, design for the environment, ISO 59000, and 9R strategies are linked with the product stewardship strategic capability (Arana-Landin et al. 2024; Morea et al. 2021; Schmitt et al. 2021; Seth and Rehman 2022; Zorpas 2024). Further, different tools, including ISO 45001, ISO 50001, DFE, EMS, Forest Stewardship Council (FSC) certification, and use of bio-based materials, are concerned with the long-term sustainability, resource conservation, employee wellbeing, and environmental protection, hence associated with the sustainable development strategic capability (Ahmed et al. 2021; Birch et al. 2012; Coca et al. 2019; Ghazilla et al. 2015; Heras-Saizarbitoria et al. 2024; Santos et al. 2021; Tiwari et al. 2020).

3.2.2 Institutional Theory-Based View of Green-Lean-Six Sigma

The institutional theory-based view (ITBV) is derived from an institutional theory, acknowledging the role of external forces—including institutions, formal and informal norms, rules, and beliefs—in shaping an organization's activities, behaviour, and strategic decisions concerning matters related to sustainability and social responsibility (Gupta and Gupta 2021; Risi et al. 2023; Sarkis et al. 2011). It emphasizes a process where business strategies and practices are successfully integrated into an organization's processes and activities as "an accepted norm in the social environment" (Caldera et al. 2019, p. 578). The ITBV focuses on the institutional isomorphisms that trigger change in an organization's existing processes (Anlesinya et al. 2023; DiMaggio and Powell 1983). It emphasizes the isomorphisms: forces and drivers behind adopting business strategies in an organization to achieve sustainability-related goals (Ain et al. 2025; Subramanian et al. 2024; Zsidisin et al. 2005). As such, the three institutional isomorphisms comprise coercive, mimetic, and normative isomorphisms. Moreover, these isomorphisms influence the adoption of strategies in an organization either individually or in combination with each other (DiMaggio and Powell 1983; Gupta and Gupta 2021; Peng and Zhang 2022).

According to DiMaggio and Powell (1983, p. 150), coercive isomorphisms refer to "both formal and informal pressures exerted on organizations by other organizations upon which they are dependent, and by cultural expectations in the society within which the organizations function". Coercive isomorphisms include the external drivers propelling organizations to implement sustainable practices, such as environmental regulations, legislation, fines, and penalties from regulatory bodies for non-compliance (Lin and Sheu 2012; Mzembe et al. 2019; Rivera 2004). It includes the pressure from customers, government, and suppliers to embrace sustainable strategies (Adebanjo et al. 2016a; Sajjad et al. 2015, 2019; Shu et al. 2024).

Normative isomorphisms, on the other hand, refer to the informal social pressure due to the increasing environmental awareness among customers, consumers, and the general public, driving firms to adopt environmentally benign practices (Ball and Craig 2010; Baumol et al. 2007; Caldera et al. 2019; Mzembe et al. 2019). It also emphasizes the pressures from social media, NGOs, and the public, as well as cultural aspects, ethical concerns, and societal expectations that play a significant role in influencing an organization to execute sustainable practices (Gupta and Barua 2017; Oelze et al. 2016; Sajjad et al. 2019).

Mimetic isomorphisms focus on a firm's efforts to imitate its competitors' best practices, operational excellence techniques, and leadership styles to achieve sustainable outcomes (Aerts et al. 2006; Farrukh and Sajjad 2024; Sarkis et al. 2011; Zhu et al. 2012). For example, competitors play a significant role in shaping an industry's dynamics in response to emerging trends, and therefore, organizations in a specific sector consider and follow key strategies adopted by their competitors in a particular domain (Caldera et al. 2019; DiMaggio and Powell 1983; Farrukh and Sajjad 2024).

In line with the above explanation, the ITBV can be used to explore the institutional pressures framing an organization's efforts to achieve economic, environmental, and social sustainability performance. In addition, it provides an appropriate organizational theoretical lens to understand the barriers and challenges to adopting sustainability practices in organizations (Caldera et al. 2019; Farrukh and Sajjad 2024). Overall, the ITBV enables examining and analysing the formal and informal factors, including regulatory requirements, cultural and social aspects, stakeholders' concerns, and competitors' influence, allowing a firm's practices to seek legitimacy (Baumol et al. 2007; Caldera et al. 2019; Roy 1997).

Aligning with the above discussion, organizations are adopting a GLSS strategy in response to the formal and informal institutional forces in the form of coercive, normative, and mimetic isomorphisms. For example, coercive isomorphisms in the form of increasing environmental regulations, such as the EU directives on Waste Electrical and Electronic Equipment (WEEE) and the EU Due Diligence Act, propel firms to adopt sustainable practices (Bening et al. 2021; Beyers et al. 2023; Laosirihongthong et al. 2013). However, the regulatory pressures vary across countries, such as the more stringent environmental regulations in developed countries compared to developing countries, which could be attributed to a lack of regulatory frameworks, policies, and monitoring frameworks (Ye et al. 2021).

Similarly, customers influence firms to adopt a sustainable business strategy, such as the GLSS, by demanding environmentally friendly products along with the aspects of improved quality, minimum price, and on-time delivery, which act as a coercive pressure (Sajjad et al. 2019; Thun and Müller 2010). Moreover, suppliers drive organizations to adopt a GLSS strategy to improve and align manufacturing processes by producing eco-friendly and sustainable materials. On the other hand, competitors encourage organizations to execute a GLSS strategy. For instance, organizations worldwide followed in the footsteps of Toyota in implementing the lean strategy and the examples of Motorola, Allied Signal, and General Motors in executing the Six Sigma strategy to achieve operational excellence (Braunscheidel et al. 2011; Dhiravidamani et al. 2017; Farrukh et al. 2021; Jasti and Kodali 2016). Similarly, several organizations in the carpet industry followed the precedent set by Interface, Inc., a leading US-based carpet manufacturer, in its commercial reuse and recycling practices (Rajala et al. 2016).

Normative isomorphism concerns informal institutions such as the World Wildlife Fund (WWF), Ellen MacArthur Foundation, and Friends of the Earth, which propel organizations to deploy environmental practices (Farrukh et al. 2021; Lauesen 2013). On the other hand, professional societies, including the International Association for Six Sigma Certification (IASSC) and the American Society for Quality (ASQ), drive organizations to execute the Six Sigma and lean strategies for improving operational performance (Braunscheidel et al. 2011; Farrukh et al. 2021).

3.2.3 Stakeholder Theory and Green-Lean-Six Sigma

According to Freeman (2010, p. 26), a stakeholder refers to "any group or individual that can affect or be affected by the realization of an organization's purpose" (Freeman 2010, p. 26). An organization's stakeholders include both the internal and external stakeholders. Internal stakeholders include employees, corporate managers, and owners (Busse et al. 2017; Clarkson 1995; Julie et al. 2024). Conversely, the external stakeholders refer to the suppliers, customers, regulatory bodies and authorities, societal groups, NGOs, local community, media, and government agencies (Busse et al. 2017; Clarkson 1995; Julie et al. 2024). While organizations influence some stakeholders, they cannot affect those in the broad environment, such as political, legal, social, and cultural forces (Freeman 2010; Harrison 1998). Moreover, the external stakeholders significantly influence the general public's opinion regarding a firm's business practices and sustainability issues (Julie et al. 2024; Silvestre 2015).

Stakeholder theory can assist practitioners and corporate managers in determining the type of relationship their organizations have with the key stakeholders to gain a competitive advantage and create value (Ahsan and Pedersen 2018; Freeman et al. 2004; Sajjad et al. 2015, 2019). From this perspective, researchers emphasize three types of stakeholders: descriptive, instrumental, and normative (Boesso et al. 2013; Donaldson and Preston 1995; Egels-Zandén and Sandberg 2010). Descriptive stakeholders refer to those parties that can be or are affected by an organization's operational processes (Ahsan and Pedersen 2018; Reed 2002). Normative stakeholders are those groups that hold a normative stance about the organization's activities (Fassin 2012; Reed 2002). Conversely, instrumental stakeholders such as political forces and regulatory authorities influence firms in designing and developing their corporate objectives (Bouguerra et al. 2023; Susnienė and Sargūnas 2011). Moreover, firms largely depend on the instrumental stakeholders to run their business operations (Ahsan and Pedersen 2018; Gallego-Alvarez et al. 2017; Saleem et al. 2020).

In addition, this theory emphasizes that an organization's success depends on the relationships with various stakeholders (Luthra et al. 2022). From the operations and supply chain management perspective, stakeholders are all parties involved in an organization's different operational processes, from raw material extraction to the products' end-of-life (Kayikci et al. 2022; Liu et al. 2025; Luthra et al. 2022; Sun et al. 2025). The interrelationships among stakeholders play a significant role in the success of an organization's strategy (Luthra et al. 2022; Wang et al. 2022). In addition, the stakeholder theory focuses on different parties' compatible and incompatible interests while examining the complex and dynamic relationships between firms and stakeholders (Luthra et al. 2022).

In terms of the relationship between the GLSS and stakeholder theory, stakeholders are concerned with both quality management (Mehralian et al. 2016; Sila 2018) and environmental protection (Awan et al. 2021; Farrukh and Sajjad 2024). For example, from the quality management perspective, the Malcolm Baldrige National Quality Award (MBNQA) framework emphasizes that stakeholders are all groups that are or might be affected by an organization's actions and success. Key stakeholders might include customers, the workforce, partners, collaborators, governing boards, stockholders, donors, suppliers, taxpayers, regulatory bodies, policy makers, funders, and local and professional communities (Sila 2018). In addition, the MNBQA suggests that an organization should follow a balanced approach in achieving results and "its strategies should not inappropriately trade off among important stakeholders, objectives, or short and longer-term goals" (Sila 2018, p. 212). This perspective of the Baldrige award is also aligned with the normative aspect of stakeholder theory that recommends managers maintain a balance while addressing multiple stakeholders' conflicting interests, including suppliers, employees, customers, government, and NGOs (Parmar et al. 2010).

From the environmental management perspective, green practices such as ISO 14001 also suggest considering the internal and external stakeholders' requirements while designing and implementing an environmental management system in the organization (Ahmed and Mathrani 2023; Ahmed et al. 2021; Boiral et al. 2018) since a firm's relationship with its stakeholders plays a significant role in creating value for an organization and achieving sustainable competitive advantage (Jones 1995; Santos et al. 2015). Additionally, the relationship between stakeholder theory and GLSS strategy can be examined by an organization's internal and external motivations for adopting a sustainable business strategy. In this sense, the stakeholder theory plays an essential role in determining a firm's collaboration and engagement with internal and external stakeholders—such as recyclers, waste collectors, NGOs, government, customers, suppliers, and consumers—that play an essential role in the success of a GLSS strategy (Farrukh et al. 2023b; Govindan and Hasanagic 2018; Kumar et al. 2016).

3.2.4 *Intellectual Capital-Based View of Green-Lean-Six Sigma*

An organization's intangible assets, such as structural, human, and relational resources, also create value for the organization and achieve competitive advantage, which refers to the intellectual capital of a firm (Asiaei et al. 2023; Costa et al. 2017; Salim et al. 2019). Several scholars have emphasized the adoption of the intellectual capital-based view (ICBV) in understanding the role played by an organization's human resources, stakeholders, systems, and procedures in addressing sustainability objectives (Farrukh et al. 2023b; Yusoff et al. 2019). Accordingly, this theoretical lens is used in domains focusing on environmental sustainability, sustainable human resource management, and business sustainability, hence it is recognized as the green ICBV (Chen 2007; Chen and Chang 2011; Yusoff et al. 2019).

With the growing environmental concerns, manufacturing organizations are trying to manage their resources in an environmentally friendly manner. However, only a few scholars have investigated organizations' intellectual capital from the environmental performance perspective and have integrated the ICBV with the natural environment (Farrukh 2025; Yusoff et al. 2019). For example, Chen et al. (2008) proposed the concept of green intellectual capital, a combination of an organization's intangible resources, knowledge, capabilities, and relationships related to environmental performance. Similarly, Huang et al. (2011) also argued that green intellectual capital facilitates organizations in creating value for the business, complying with environmental regulations, and fulfilling the customers' environmental requirements.

While the NRBV addresses the GLSS tools and practices aligned with pollution prevention, product stewardship, and sustainable development strategic capabilities, the success of this strategy also depends on various other factors that play a significant role in its deployment. Accordingly, the ICBV can help people to understand the contribution of human resources, stakeholders, and the organization's systems in executing the GLSS approach (Farrukh et al. 2023b). For example, an organization's structural capital, including information technology infrastructure, standard operating procedures, organizational culture, and the role of top management, is among the critical success factors (CSFs) of a GLSS execution (Chen 2007). In addition, an organization's green structural capital can also be beneficial in maintaining the employees' environmental knowledge even when they leave the firm. For example, adequate information system and technological resources (such as cloud databases) can help managers store environmental knowledge and information obtained from both internal and external stakeholders. Hence, organizations need to invest in establishing and developing structural capital as a critical aspect of green intellectual capital (Chang and Chen 2012; Yousefi and Rodriguez Hernandez 2019).

Similarly, green human capital, such as skills, commitment, expertise, attitude, and knowledge of an organization's employees, play a significant role in implementing various GLSS practices to achieve sustainable competitive advantage (Chen 2007; Yusoff et al. 2019). On the other hand, an organization's relational capital, comprising the organization's connections and links with customers, suppliers, recyclers, waste collectors, and institutions, helps in executing GLSS tools for enhancing environmental performance (Chen 2007). In this regard, an organization's strong relationship and networking with the suppliers, governments, customers, investors, shareholders, and institutions can assist in increasing environmental awareness among stakeholders, disseminating information and sharing knowledge, and achieving environmental sustainability (Dickel et al. 2018; Yusoff et al. 2019).

In conclusion, this chapter discussed different organizational theories used in the extant literature in the GLSS domain, including the NRBV, ITBV, stakeholder theory, and ICBV. Further, the relationship between various organizational theories and the GLSS constructs, including barriers, drivers, tools, practices, and success factors, has been presented, which can help in understanding the theoretical alignment for promoting the GLSS strategy.

3.3 Summary

This chapter examined the leading organizational perspectives that explain GLSS adoption and implementation. It argued that overarching organizational theories are central to understanding key strategies and approaches in the business context. To this end, theoretical lenses, including the natural resource-based, intellectual capital-based, institutional, and stakeholder theories, were discussed critically, considering emerging research work. This rich, theory-informed discussion provides a deeper understanding and theoretical justification for companies to embrace the GLSS approach.

References

Abobakr, M. A., Abdel-Kader, M., & Elbayoumi, A. F. (2023). Integrating S-ERP systems and lean manufacturing practices to improve sustainability performance: an institutional theory perspective. *Journal of Accounting in Emerging Economies, 13*(5), 870–897.

Abobakr, M. A., Abdel-Kader, M. G., & Elbayoumi, A. F. (2022). The impact of lean manufacturing practices on sustainability performance: A natural resource-based view. *Journal of Modern Accounting and Auditing, 18*(3), 115–130.

Adebanjo, D., Samaranayake, P., Mafakheri, F., & Laosirihongthong, T. (2016a). Prioritization of Six-Sigma project selection. *Benchmarking: An International Journal, 23*(7), 1983–2003. https://doi.org/10.1108/bij-09-2015-0086

Adebanjo, D., Samaranayake, P., Mafakheri, F., & Laosirihongthong, T. (2016b). Prioritization of Six-Sigma project selection: a resource-based view and institutional norms perspective. *Benchmarking: An International Journal, 23*(7), 1983–2003.

Aerts, W., Cormier, D., & Magnan, M. (2006). Intra-industry imitation in corporate environmental reporting: An international perspective. *Journal of Accounting and public Policy, 25*(3), 299–331.

Afum, E., Agyabeng-Mensah, Y., Baah, C., Asamoah, G., & Yaw Kusi, L. (2023). Green market orientation, green value-based innovation, green reputation and enterprise social performance of Ghanaian SMEs: the role of lean management. *Journal of Business & Industrial Marketing, 38*(10), 2151–2169.

Afum, E., Li, Y., Han, P., & Sun, Z. (2022). Interplay between lean management and circular production system: implications for zero-waste performance, green value competitiveness, and social reputation. *Journal of Manufacturing Technology Management, 33*(7), 1213–1231.

Agyabeng-Mensah, Y., Afum, E., Acquah, I. S. K., Dacosta, E., Baah, C., & Ahenkorah, E. (2021). The role of green logistics management practices, supply chain traceability and logistics ecocentricity in sustainability performance. *The International Journal of Logistics Management, 32*(2), 538–566.

Ahmad, I., Abdullah, A., Khalik, A., & Putra, A. H. P. K. (2025). The mediating role of green accounting management on financial performance: integrated stakeholder theory and natural resource-based view. *International Journal of Energy Economics and Policy, 15.*

Ahmad, S., Abdullah, A., & Talib, F. (2021). Lean-green performance management in Indian SMEs: a novel perspective using the best-worst method approach. *Benchmarking: An International Journal*, *28*(2), 737–765.

Ahmed, A., & Mathrani, S. (2023). Critical success factors for a combined lean and ISO 14001 implementation in the manufacturing industry: a systematic literature review. *The TQM Journal*, *36*(7), 2071–2106. https://doi.org/10.1108/tqm-12-2022-0347

Ahmed, A., Mathrani, S., & Jayamaha, N. (2021). An integrated lean and ISO 14001 framework for environmental performance: an assessment of New Zealand meat industry. *International Journal of Lean Six Sigma*, *ahead-of-print*(ahead-of-print). https://doi.org/10.1108/ijlss-05-2021-0100

Ahsan, D., & Pedersen, S. (2018). The influence of stakeholder groups in operation and maintenance services of offshore wind farms: Lesson from Denmark. *Renewable Energy*, *125*, 819–828. https://doi.org/10.1016/j.renene.2017.12.098

Ain, Q. U., Javaid, H. M., Cafaro, E. M., & D'Alessio, R. (2025). Steering Towards Green: Do Technology Directors Matter for Green Innovation? An Institutional Isomorphism Approach. *Corporate Social Responsibility and Environmental Management*.

Al-Hakimi, M. A., Al Koliby, I. S., Al-Swidi, A. K., Mehat, N. A. B., & Taleb, T. S. T. (2025). From quality to green innovation: the catalytic role of green organizational culture in TQM and green manufacturing practice. *International Journal of Innovation Science*.

Alkaraan, F., Elmarzouky, M., Hussainey, K., Venkatesh, V. G., Shi, Y., & Gulko, N. (2024). Reinforcing green business strategies with Industry 4.0 and governance towards sustainability: Natural-resource-based view and dynamic capability. *Business Strategy and the Environment*, *33*(4), 3588–3606.

Anlesinya, A., Amponsah-Tawiah, K., Dartey-Baah, K., Adeti, S. K., & Brefo-Manuh, A. B. (2023). Institutional isomorphism and sustainable HRM adoption: a conceptual framework. *Industrial and Commercial Training*, *55*(1), 62–76.

Arana-Landin, G., Sigüenza, W., Landeta-Manzano, B., & Laskurain-Iturbe, I. (2024). Circular economy: On the road to ISO 59000 family of standards. *Corporate Social Responsibility and Environmental Management*, *31*(3), 1977–2009.

Asiaei, K., O'Connor, N. G., Barani, O., & Joshi, M. (2023). Green intellectual capital and ambidextrous green innovation: The impact on environmental performance. *Business Strategy and the Environment*, *32*(1), 369–386.

Awan, U., Sroufe, R., & Shahbaz, M. (2021). Industry 4.0 and the circular economy: A literature review and recommendations for future research. *Business Strategy and the Environment*, *30*(4), 2038–2060.

Ball, A., & Craig, R. (2010). Using neo-institutionalism to advance social and environmental accounting. *Critical Perspectives on Accounting*, *21*(4), 283–293. https://doi.org/10.1016/j.cpa.2009.11.006

Barney, J. (1991). Firm resources and sustained competitive advantage. *Journal of management*, *17*(1), 99–120.

Baumgartner, R. J., & Rauter, R. (2017). Strategic perspectives of corporate sustainability management to develop a sustainable organization. *Journal of Cleaner Production*, *140*, 81–92. https://doi.org/10.1016/j.jclepro.2016.04.146

Baumol, W. J., Litan, R. E., & Schramm, C. J. (2007). *Good capitalism, bad capitalism, and the economics of growth and prosperity*. Yale University Press.

Beamish, P. W., & Chakravarty, D. (2021). Using the resource-based view in multinational enterprise research. *Journal of Management*, *47*(7), 1861–1877.

Bening, C. R., Pruess, J. T., & Blum, N. U. (2021). Towards a circular plastics economy: Interacting barriers and contested solutions for flexible packaging recycling. *Journal of Cleaner Production*, *302*. https://doi.org/10.1016/j.jclepro.2021.126966

Beyers, F., Leventon, J., & Heinrichs, H. (2023). Collaborative governance or state regulation? Endless efforts but little capacity for sustainability transformation of the German textile sector. *Environmental Policy and Governance*, *33*(1), 56–77.

Birch, A., Hon, K., & Short, T. (2012). Structure and output mechanisms in Design for Environment (DfE) tools. *Journal of Cleaner Production*, *35*, 50–58. https://doi.org/10.1016/j.jclepro.2012.05.029

Boesso, G., Kumar, K., & Michelon, G. (2013). Descriptive, instrumental and strategic approaches to corporate social responsibility: Do they drive the financial performance of companies differently? *Accounting, Auditing & Accountability Journal*, *26*(3), 399–422.

Boiral, O., Guillaumie, L., Heras-Saizarbitoria, I., & Tayo Tene, C. V. (2018). Adoption and Outcomes of ISO 14001: A Systematic Review. *International Journal of Management Reviews*, *20*(2), 411–432. https://doi.org/10.1111/ijmr.12139

Bouguerra, A., Hughes, M., Cakir, M. S., & Tatoglu, E. (2023). Linking entrepreneurial orientation to environmental collaboration: a stakeholder theory and evidence from multinational companies in an emerging market. *British journal of management*, *34*(1), 487–511.

Braunscheidel, M. J., Hamister, J. W., Suresh, N. C., & Star, H. (2011). An institutional theory perspective on Six Sigma adoption. *International Journal of Operations & Production Management*, *31*(4), 423–451. https://doi.org/10.1108/01443571111119542

Buhaya, M. I., & Metwally, A. B. M. (2024). Green intellectual capital and green supply chain performance: do external pressures matter? *Cogent Business & Management*, *11*(1), 2349276.

Busse, C., Regelmann, A., Chithambaram, H., & Wagner, S. M. (2017). Managerial perceptions of energy in logistics. *International Journal of Physical Distribution & Logistics Management*, *47*(6), 447–471. https://doi.org/10.1108/ijpdlm-04-2015-0090

Caldera, H. T. S., Desha, C., & Dawes, L. (2019). Evaluating the enablers and barriers for successful implementation of sustainable business practice in 'lean' SMEs. *Journal of Cleaner Production*, *218*, 575–590. https://doi.org/10.1016/j.jclepro.2019.01.239

Chang, C. H., & Chen, Y. S. (2012). The determinants of green intellectual capital. *Management Decision*, *50*(1), 74–94.

Chang, R.-D., Zuo, J., Zhao, Z.-Y., Zillante, G., Gan, X.-L., & Soebarto, V. (2017). Evolving theories of sustainability and firms: History, future directions and implications for renewable energy research. *Renewable and Sustainable Energy Reviews*, *72*, 48–56.

Chen, J. C., Li, Y., & Shady, B. D. (2008). From value stream mapping toward a lean/sigma continuous improvement process: an industrial case study. *International Journal of Production Research*, *48*(4), 1069–1086. https://doi.org/10.1080/00207540802484911

Chen, Y.-S. (2007). The Positive Effect of Green Intellectual Capital on Competitive Advantages of Firms. *Journal of Business Ethics*, *77*(3), 271–286. https://doi.org/10.1007/s10551-006-9349-1

Chen, Y.-S., & Chang, C.-H. (2011). Enhance environmental commitments and green intangible assets toward green competitive advantages: an analysis of structural equation modeling (SEM). *Quality & Quantity*, *47*(1), 529–543. https://doi.org/10.1007/s11135-011-9535-9

Chiarini, A. (2014). Sustainable manufacturing-greening processes using specific Lean Production tools: an empirical observation from European motorcycle component manufacturers. *Journal of Cleaner Production*, *85*, 226–233. https://doi.org/10.1016/j.jclepro.2014.07.080

Clarkson, M. E. (1995). A stakeholder framework for analyzing and evaluating corporate social performance. *Academy of management review*, *20*(1), 92–117.

Cobb, J., & Daly, H. (1989). For the common good, redirecting the economy toward community, the environment and a sustainable future. In: Boston: Beacon Press.

Coca, G., Castrillón, O. D., Ruiz, S., Mateo-Sanz, J. M., & Jiménez, L. (2019). Sustainable evaluation of environmental and occupational risks scheduling flexible job shop manufacturing systems. *Journal of Cleaner Production*, *209*, 146–168. https://doi.org/10.1016/j.jclepro.2018.10.193

Coppola, C., Vollero, A., & Siano, A. (2023). Developing dynamic capabilities for the circular economy in the textile and clothing industry in Italy: A natural-resource-based view. *Business Strategy and the Environment*, *32*(7), 4798–4820.

Costa, R. V., Fernández-Jardon Fernández, C., & Figueroa Dorrego, P. (2017). Critical elements for product innovation at Portuguese innovative SMEs: an intellectual capital perspective. *Knowledge Management Research & Practice*, *12*(3), 322–338. https://doi.org/10.1057/kmrp.2014.15

Cristina De Stefano, M., Montes-Sancho, M. J., & Busch, T. (2016). A natural resource-based view of climate change: Innovation challenges in the automobile industry. *Journal of Cleaner Production, 139*, 1436–1448. https://doi.org/10.1016/j.jclepro.2016.08.023

Dhiravidamani, P., Ramkumar, A. S., Ponnambalam, S. G., & Subramanian, N. (2017). Implementation of lean manufacturing and lean audit system in an auto parts manufacturing industry—an industrial case study. *International Journal of Computer Integrated Manufacturing, 31*(6), 579–594. https://doi.org/10.1080/0951192x.2017.1356473

Dickel, P., Hörisch, J., & Ritter, T. (2018). Networking for the environment: The impact of environmental orientation on start-ups' networking frequency and network size. *Journal of Cleaner Production, 179*, 308-316.

DiMaggio, P. J., & Powell, W. W. (1983). The iron cage revisited: Institutional isomorphism and collective rationality in organizational fields. *American sociological review*, 147–160.

Donaldson, T., & Preston, L. E. (1995). The stakeholder theory of the corporation: Concepts, evidence, and implications. *Academy of management review, 20*(1), 65–91.

Egels-Zandén, N., & Sandberg, J. (2010). Distinctions in descriptive and instrumental stakeholder theory: A challenge for empirical research. *Business Ethics: A European Review, 19*(1), 35–49.

Farrukh, A. (2022). A green-lean-six sigma model for environmental performance in manufacturing organizations: a study of a developed and developing nation: a thesis presented in partial fulfilment of the requirement for the degree of Doctor of Philosophy in Engineering Management, School of Food and Advanced Technology, Massey University, Auckland, New Zealand.

Farrukh, A. (2025). A comparative exploratory study of the critical success factors of a green-lean-six sigma strategy. *Business process Management journal*, 1–35. https://doi.org/10.1108/BPMJ-09-2024-0862

Farrukh, A., Mathrani, S., & Sajjad, A. (2021). A natural resource and institutional theory-based view of green-lean-six sigma drivers for environmental management. *Business Strategy and the Environment, 31*(3), 1074–1090. https://doi.org/10.1002/bse.2936

Farrukh, A., Mathrani, S., & Sajjad, A. (2022). Managerial perspectives on green-lean-six sigma adoption in the flexible packaging industry: empirical evidence from an emerging economy. *Journal of Manufacturing Technology Management, ahead-of-print*(ahead-of-print). https://doi.org/10.1108/JMTM-02-2022-0080

Farrukh, A., Mathrani, S., & Sajjad, A. (2023a). An exploratory study of green-lean-six sigma motivators for environmental sustainability: Managerial insights from a developed and developing economy. *Business Strategy and the Environment*. https://doi.org/10.1002/bse.3412

Farrukh, A., Mathrani, S., & Sajjad, A. (2023b). Green-lean-six sigma practices and supporting factors for transitioning towards circular economy: A natural resource and intellectual capital-based view. *Resources Policy, 84*. https://doi.org/10.1016/j.resourpol.2023.103789

Farrukh, A., & Sajjad, A. (2024). Drivers for and barriers to circular economy transition in the textile industry: A developing economy perspective. *Sustainable Development*. https://doi.org/10.1002/sd.3088

Fassin, Y. (2012). Stakeholder management, reciprocity and stakeholder responsibility. *Journal of Business Ethics, 109*(1), 83–96.

Ferreira, N. C., & Ferreira, J. J. M. (2025). The field of resource-based view research: mapping past, present and future trends. *Management Decision, 63*(4), 1124–1153.

Fiksel, J. (1993). Design for environment: the new quality imperative. *Corporate environmental strategy, 1*(3), 49–55.

Foulon, B., & Marsat, S. (2023). Does environmental footprint influence the resilience of firms facing environmental penalties? *Business Strategy and the Environment, 32*(8), 6154–6168.

Freeman, R. E. (2010). *Stakeholder theory: The state of the art*. Cambridge University Press.

Freeman, R. E., Wicks, A. C., & Parmar, B. (2004). Stakeholder theory and "the corporate objective revisited". *Organization science, 15*(3), 364–369.

Gabriel, C.-A., Bortsie-Aryee, N. A., Apparicio-Farrell, N., & Farrell, E. (2018). How supply chain choices affect the life cycle impacts of medical products. *Journal of Cleaner Production, 182*, 1095–1106. https://doi.org/10.1016/j.jclepro.2018.02.107

Gallego-Alvarez, I., Ortas, E., Vicente-Villardón, J. L., & Alvarez Etxeberria, I. (2017). Institutional constraints, stakeholder pressure and corporate environmental reporting policies. *Business Strategy and the Environment*, *26*(6), 807–825.

Garza-Reyes, J. A., Kumar, V., Chaikittisilp, S., & Tan, K. H. (2018). The effect of lean methods and tools on the environmental performance of manufacturing organisations. *Int. J. Prod. Econ.*, *200*, 170–180. https://doi.org/10.1016/j.ijpe.2018.03.030

Ghazilla, R. A. R., Sakundarini, N., Abdul-Rashid, S. H., Ayub, N. S., Olugu, E. U., & Musa, S. N. (2015). Drivers and Barriers Analysis for Green Manufacturing Practices in Malaysian SMEs: A Preliminary Findings. *Procedia CIRP*, *26*, 658–663. https://doi.org/10.1016/j.procir.2015.02.085

Govindan, K., & Hasanagic, M. (2018). A systematic review on drivers, barriers, and practices towards circular economy: a supply chain perspective. *International Journal of Production Research*, *56*(1–2), 278–311.

Grama-Vigouroux, S., Gull, A. A., & Rind, A. A. (2025). Examining the role of eco-innovation in mitigating greenhouse gas emissions: A natural resource-based view. *Journal of Cleaner Production*, 145714.

Gueler, M. S., & Schneider, S. (2021). The resource-based view in business ecosystems: A perspective on the determinants of a valuable resource and capability. *Journal of Business Research*, *133*, 158–169.

Gupta, A. K., & Gupta, N. (2021). Environment practices mediating the environmental compliance and firm performance: An institutional theory perspective from emerging economies. *Global Journal of Flexible Systems Management*, *22*(3), 157–178.

Gupta, H., & Barua, M. K. (2017). Supplier selection among SMEs on the basis of their green innovation ability using BWM and fuzzy TOPSIS. *Journal of Cleaner Production*, *152*, 242–258. https://doi.org/10.1016/j.jclepro.2017.03.125

Hamdoun, M. (2020). The antecedents and outcomes of environmental management based on the resource-based view: A systematic literature review. *Management of Environmental Quality: An International Journal*, *31*(2), 451–469.

Harrison, J. S. (1998). *Strategic Management of Organizations and Stakeholders: Cases*. South-Western College Publishing.

Hart, S. L. (1995). A natural-resource-based view of the firm. *Academy of management review*, *20*(4), 986–1014.

Hart, S. L., Barney, J. B., Ketchen, D. J., Wright, M., & Dowell, G. (2010). Invited Editorial: A Natural-Resource-Based View of the Firm. *Journal of Management*, *37*(5), 1464–1479. https://doi.org/10.1177/0149206310390219

Heras-Saizarbitoria, I., Boiral, O., & Testa, F. (2024). ISO 59000 standards for the Circular Economy: a call for accuracy. *Circular Economy and Sustainability*, 1–7.

Huang, Y. C., Wu, Y. C. J., Wang, Y. C., & Boulanger, N. C. (2011). Decision making in online auctions. *Management Decision.*

Jasti, N. V. K., & Kodali, R. (2016). An empirical study for implementation of lean principles in Indian manufacturing industry. *Benchmarking: An International Journal*, *23*(1), 183–207. https://doi.org/10.1108/bij-11-2013-0101

Johnson-Hall, T. D., & Hall, D. C. (2022). Redefining quality in food supply chains via the natural resource based view and convention theory. *Sustainability*, *14*(15), 9456.

Jones, T. M. (1995). Instrumental stakeholder theory: A synthesis of ethics and economics. *Academy of management review*, *20*(2), 404–437.

Julie, S., Potter, A., & Geng, R. (2024). Examining the effects of stakeholder forces on sustainable practices in the Bangladeshi garment industry. *Cleaner Logistics and Supply Chain*, *12*. https://doi.org/10.1016/j.clscn.2024.100162

Kayikci, Y., Kazancoglu, Y., Gozacan-Chase, N., & Lafci, C. (2022). Analyzing the drivers of smart sustainable circular supply chain for sustainable development goals through stakeholder theory. *Business Strategy and the Environment*, *31*(7), 3335–3353.

Keoleian, G. A. (1993). Life Cycle Design Guidance Manual. *United States Environmental Protection Agency publications.*

Khanra, S., Kaur, P., Joseph, R. P., Malik, A., & Dhir, A. (2022). A resource-based view of green innovation as a strategic firm resource: Present status and future directions. *Business Strategy and the Environment, 31*(4), 1395–1413.

Kleiner, A. (1991). What does it mean to be green? *Harvard business review, 69*(4), 38–42.

Kühl, C., Bourlakis, M., Aktas, E., & Skipworth, H. (2022). Product-service systems and circular supply chain practices in UK SMEs: The moderating effect of internal environmental orientation. *Journal of Business Research, 146*, 155–165.

Kumar, S., Luthra, S., Govindan, K., Kumar, N., & Haleem, A. (2016). Barriers in green lean six sigma product development process: an ISM approach. *Production Planning & Control*, 1–17. https://doi.org/10.1080/09537287.2016.1165307

Laosirihongthong, T., Adebanjo, D., & Choon Tan, K. (2013). Green supply chain management practices and performance. *Industrial Management & Data Systems, 113*(8), 1088–1109.

Lau, C. C. I., & Wong, C. W. Y. (2024). Achieving sustainable development with sustainable packaging: A natural-resource-based view perspective. *Business Strategy and the Environment, 33*(5), 4766–4787.

Lauesen, L. M. (2013). Environmental Protection Agencies (All Countries). In S. O. Idowu, N. Capaldi, L. Zu, & A. D. Gupta (Eds.), *Encyclopedia of Corporate Social Responsibility* (pp. 1006–1016). Springer Berlin Heidelberg. https://doi.org/10.1007/978-3-642-28036-8_385

Lichtenthaler, U. (2022). Explicating a sustainability-based view of sustainable competitive advantage. *Journal of strategy and management, 15*(1), 76–95.

Lin, R.-j., & Sheu, C. (2012). Why do firms adopt/implement green practices?–an institutional theory perspective. *Procedia-Social and Behavioral Sciences, 57*, 533–540.

Liu, Q., Dong, Y., Zhao, M., He, X., & Zhang, X. (2025). How corporate environmental responsibility enhances sustainable performance through green supply chain management under stakeholder theory perspective. *Asia Pacific Journal of Marketing and Logistics*, 1–18.

Liu, S. (2025). Green management promotes long-term business competitive advantage through the resource-based view. *Total Quality Management & Business Excellence*, 1–28.

Llach, J., Palau-Pinyana, E., Lei, L., & Perramon, J. (2025). Key enablers for energy firms in implementing the SDGs: Lessons based on a resource-based view approach. *Technological Forecasting and Social Change, 213*, 124011.

Luthra, S., Sharma, M., Kumar, A., Joshi, S., Collins, E., & Mangla, S. (2022). Overcoming barriers to cross-sector collaboration in circular supply chain management: a multi-method approach. *Transportation Research Part E: Logistics and Transportation Review, 157*. https://doi.org/10.1016/j.tre.2021.102582

Mehralian, G., Nazari, J. A., Zarei, L., & Rasekh, H. R. (2016). The effects of corporate social responsibility on organizational performance in the Iranian pharmaceutical industry: The mediating role of TQM. *Journal of Cleaner Production, 135*, 689–698.

Mengistu, A. T., Dieste, M., Panizzolo, R., & Biazzo, S. (2025). Achieving Product Sustainability: Analyzing Key Factors Influencing Sustainable Product Design in Manufacturing Companies. *Corporate Social Responsibility and Environmental Management.*

Menguc, B., & Ozanne, L. K. (2005). Challenges of the "green imperative": A natural resource-based approach to the environmental orientation–business performance relationship. *Journal of Business Research, 58*(4), 430–438.

Mohamad, E., Ishak, A., & Arep, H. (2024). Promoting sustainable wastewater treatment operations: The role of lean practices and lean six sigma enablers. *Journal of Mechanical Engineering and Sciences*, 10116–10126.

Morea, D., Fortunati, S., & Martiniello, L. (2021). Circular economy and corporate social responsibility: Towards an integrated strategic approach in the multinational cosmetics industry. *Journal of Cleaner Production, 315*. https://doi.org/10.1016/j.jclepro.2021.128232

Münch, C., Benz, L. A., & Hartmann, E. (2022). Exploring the circular economy paradigm: A natural resource-based view on supplier selection criteria. *Journal of Purchasing and Supply Management*, *28*(4), 100793.

Mzembe, A. N., Melissen, F., & Novakovic, Y. (2019). Greening the hospitality industry in the developing world: Analysis of the drivers and barriers. *Business Ethics: A European Review*, *28*(3), 335–348.

Nayak, B., Bhattacharyya, S. S., & Krishnamoorthy, B. (2023). Integrating the dialectic perspectives of resource-based view and industrial organization theory for competitive advantage–a review and research agenda. *Journal of Business & Industrial Marketing*, *38*(3), 656–679.

Oelze, N., Hoejmose, S. U., Habisch, A., & Millington, A. (2016). Sustainable development in supply chain management: The role of organizational learning for policy implementation. *Business Strategy and the Environment*, *25*(4), 241–260.

Parmar, B. L., Freeman, R. E., Harrison, J. S., Wicks, A. C., Purnell, L., & De Colle, S. (2010). Stakeholder theory: The state of the art. *Academy of Management Annals*, *4*(1), 403–445.

Peng, X., & Zhang, R. (2022). Corporate governance, environmental sustainability performance, and normative isomorphic force of national culture. *Environmental Science and Pollution Research*, *29*(22), 33443–33473.

Rajala, R., Westerlund, M., & Lampikoski, T. (2016). Environmental sustainability in industrial manufacturing: re-examining the greening of Interface's business model. *Journal of Cleaner Production*, *115*, 52–61. https://doi.org/10.1016/j.jclepro.2015.12.057

Reed, D. (2002). Employing normative stakeholder theory in developing countries: A critical theory perspective. *Business & Society*, *41*(2), 166–207.

Risi, D., Vigneau, L., Bohn, S., & Wickert, C. (2023). Institutional theory-based research on corporate social responsibility: Bringing values back in. *International Journal of Management Reviews*, *25*(1), 3–23.

Rivera, J. (2004). Institutional pressures and voluntary environmental behavior in developing countries: Evidence from the Costa Rican hotel industry. *Society and Natural Resources*, *17*(9), 779–797.

Rooney, C. (1993). Economics of pollution prevention = how Wmte reduction pays. *Pollution Prevention Review*, *261*.

Roy, W. G. (1997). Socializing capital: The rise of the large industrial corporation in America.

Sajjad, A., Eweje, G., & Tappin, D. (2015). Sustainable Supply Chain Management: Motivators and Barriers. *Business Strategy and the Environment*, *24*(7), 643–655. https://doi.org/10.1002/bse.1898

Sajjad, A., Eweje, G., & Tappin, D. (2019). Managerial perspectives on drivers for and barriers to sustainable supply chain management implementation: Evidence from New Zealand. *Business Strategy and the Environment*, *29*(2), 592–604. https://doi.org/10.1002/bse.2389

Saleem, F., Zhang-Zhang, Y., Malik, M. I., & Allui, A. (2020). Revisiting stakeholder theory and environmentalism: Evidence from an emerging economy. *Sustainability*, *12*(20), 8751.

Salim, N., Ab Rahman, M. N., & Abd Wahab, D. (2019). A systematic literature review of internal capabilities for enhancing eco-innovation performance of manufacturing firms. *Journal of Cleaner Production*, *209*, 1445–1460. https://doi.org/10.1016/j.jclepro.2018.11.105

Santos, Â. R. S., Melo, R. M. d., Clemente, T. R. N., & Machado Santos, S. (2021). Integrated management system: methodology for maturity assessment in food industries. *Benchmarking: An International Journal*, *29*(6), 1757–1780. https://doi.org/10.1108/bij-05-2021-0280

Santos, G., Rebelo, M., Lopes, N., Alves, M. R., & Silva, R. (2015). Implementing and certifying ISO 14001 in Portugal: motives, difficulties and benefits after ISO 9001 certification. *Total Quality Management & Business Excellence*, *27*(11–12), 1211–1223. https://doi.org/10.1080/14783363.2015.1065176

Sarkis, J., Zhu, Q., & Lai, K.-h. (2011). An organizational theoretic review of green supply chain management literature. *International Journal of Production Economics*, *130*(1), 1–15.

Schmitt, T., Wolf, C., Lennerfors, T. T., & Okwir, S. (2021). Beyond "Leanear" production: A multi-level approach for achieving circularity in a lean manufacturing context. *Journal of Cleaner Production, 318*. https://doi.org/10.1016/j.jclepro.2021.128531

Sehnem, S., Bispo, D. S., João, J. O., de Souza, M. A. L., Bertoglio, O., Ciotti, R., & Deon, S. M. (2022). Upscaling circular economy in foodtechs businesses in emergent countries: Towards sustainable development through natural resource based view. *Sustainable Development, 30*(5), 1200–1221.

Seth, D., & Rehman, M. A. A. (2022). Critical success factors-based strategy to facilitate green manufacturing for responsible business: An application experience in Indian context. *Business Strategy and the Environment, 31*(7), 2786–2806. https://doi.org/10.1002/bse.3047

Sharma, S., Singh, R. K., Mishra, R., & Subramanian, N. (2024). Developing climate neutrality among supply chain members in metal and mining industry: natural resource-based view perspective. *The International Journal of Logistics Management, 35*(3), 804–832.

Shu, Y., Zhuang, X., Ying, R., & Xu, G. (2024). Formal institutional pressure and the integration of corporate environmental and financial performance: Empirical evidence from listed companies in heavily polluting industries in China. *Sustainability, 16*(6), 2471.

Sila, I. (2018). Investigating changes in TQM's effects on corporate social performance and financial performance over time. *Total Quality Management & Business Excellence, 31*(1–2), 210–229. https://doi.org/10.1080/14783363.2018.1458609

Silvestre, B. S. (2015). A hard nut to crack! Implementing supply chain sustainability in an emerging economy. *Journal of Cleaner Production, 96*, 171–181.

Skalli, D., Charkaoui, A., Cherrafi, A., Garza-Reyes, J. A., Antony, J., & Shokri, A. (2024). Analyzing the integrated effect of circular economy, Lean Six Sigma, and Industry 4.0 on sustainable manufacturing performance from a practice-based view perspective. *Business Strategy and the Environment, 33*(2), 1208–1226.

Smart, B. (1992). Beyond compliance-a new industry view of the environment.

Stalk Jr, G., & Hout, T. M. (1990). Competing against time. *Research-Technology Management, 33*(2), 19–24.

Subramanian, B., Jain, N. K., & Bhattacharyya, S. S. (2024). Exploring the direct impact of environmental, social and governance (ESG) factors on organisational innovation output and institutional isomorphism: a study in the context of multinational life sciences organisations. *Benchmarking: An International Journal.*

Sun, F., Qu, Z., Wu, B., & Bold, S. (2025). Comparative analysis of international environmental policies and supply chain sustainability. *Journal of Environmental Management, 390*, 126324.

Susnienė, D., & Sargūnas, G. (2011). Stakeholder management paradoxes from the perspective of normative, descriptive and instrumental approach.

Tan, K., Siddik, A. B., Sobhani, F. A., Hamayun, M., & Masukujjaman, M. (2022). Do environmental strategy and awareness improve firms' environmental and financial performance? The role of competitive advantage. *Sustainability, 14*(17), 10600.

Thun, J. H., & Müller, A. (2010). An empirical analysis of green supply chain management in the German automotive industry. *Business Strategy and the Environment, 19*(2), 119–132.

Tian, J., Dong, Y., Vagnani, G., & Liu, P. (2023). Green innovation and the stock market value of heavily polluting firms: The role of environmental compliance costs and technological collaboration. *Business Strategy and the Environment, 32*(7), 4938–4953.

Tiwari, P., Sadeghi, J. K., & Eseonu, C. (2020). A sustainable lean production framework with a case implementation: Practice-based view theory. *Journal of Cleaner Production, 277*, 123078.

Wang, J. X., Burke, H., & Zhang, A. (2022). Overcoming barriers to circular product design. *International Journal of Production Economics, 243*, 108346.

Ye, F., Huang, G., Zhan, Y., & Li, Y. (2021). Factors Mediating and Moderating the Relationships Between Green Practice and Environmental Performance: Buyer–Supplier Relation and Institutional Context. *IEEE Transactions on Engineering Management*, 1–14. https://doi.org/10.1109/tem.2021.3060434

Yousefi, A., & Rodriguez Hernandez, M. (2019). Using a system theory based method (STAMP) for hazard analysis in process industry. *Journal of Loss Prevention in the Process Industries*, *61*, 305–324. https://doi.org/10.1016/j.jlp.2019.06.014

Yuan, M., Wang, X., Lin, H., Wu, H., Yu, M., & Chen, X. (2023). Crafting enviropreneurial marketing through green innovation: A natural resource-based view. *IEEE Transactions on Engineering Management*, *71*, 4548–4557.

Yusoff, Y. M., Omar, M. K., Kamarul Zaman, M. D., & Samad, S. (2019). Do all elements of green intellectual capital contribute toward business sustainability? Evidence from the Malaysian context using the Partial Least Squares method. *Journal of Cleaner Production*, *234*, 626–637. https://doi.org/10.1016/j.jclepro.2019.06.153

Zhu, Q., Sarkis, J., & Lai, K.-h. (2012). Examining the effects of green supply chain management practices and their mediations on performance improvements. *International Journal of Production Research*, *50*(5), 1377–1394. https://doi.org/10.1080/00207543.2011.571937

Zorpas, A. A. (2024). The hidden concept and the beauty of multiple "R" in the framework of waste strategies development reflecting to circular economy principles. *Science of the total environment*, *952*, 175508.

Zsidisin*?, G. A., Melnyk, S. A., & Ragatz, G. L. (2005). An institutional theory perspective of business continuity planning for purchasing and supply management. *International Journal of Production Research*, *43*(16), 3401–3420.

Chapter 4
Green-Lean-Six Sigma Applications in Manufacturing and Services

4.1 Introduction

This chapter presents the application of the GLSS strategy in a range of organizations in the manufacturing and service sectors. In particular, various cases of GLSS implementation in the discrete and process industrial sectors have been critically examined and discussed within the manufacturing sector, followed by cases in the service sector, including the healthcare and higher education sectors. In addition, this chapter also demonstrates GLSS execution in Small and Medium-sized Enterprises (SMEs).

4.2 Application of Green-Lean-Six Sigma—Cases in Manufacturing and Service Sectors

4.2.1 Green-Lean-Six Sigma in Manufacturing Organizations

Several studies have been conducted in the domain of GLSS, investigating the successful implementation of GLSS in manufacturing organizations under discrete and process industries (Bhamu et al. 2013; Deepak and Dharmendra 2025; McGlinchey et al. 2025; Prasad et al. 2020; Sasikumar and Acharya 2020; Sharma and Singh 2025). The following section covers GLSS application in the discrete manufacturing sector.

A. Farrukh and A. Sajjad, *A Manager's Guide to Green-Lean-Six Sigma Application*,
https://doi.org/10.1007/978-3-032-21111-8_4

4.2.1.1 Cases of Green-Lean-Six Sigma Application in Discrete Industries

Firms operating in the discrete manufacturing industry use different machines and components to manufacture various products, such as furniture and automobiles (Gholami et al. 2020; Liu and Miao 2006; Vinnari 2024). The environmental sustainability issues of the organizations in the discrete manufacturing industry include the discharge of wastewater and effluents, the release of GHG emissions, and the use of hazardous materials (Farrukh 2022; Jin et al. 2024; Yadav et al. 2021). For example, the automobile industry faces environmental issues such as the release of fumes, leakage of oil, use of chemicals, and extensive consumption of energy (Chiarini 2014; Cristina De Stefano et al. 2016; Maldonado-Guzmán et al. 2020).

A review of the emerging literature revealed various studies investigating GLSS applications in discrete manufacturing industries. For example, a study conducted in an electrical components manufacturing organization in Gujarat, India, highlighted the enablers for GLSS in the discrete manufacturing sector (Parmar and Desai 2020). The case organization was an ISO 14001:2015 and ISO 9001:2015 certified company and produced a variety of electrical products for the Indian railways, including bus bars, terminal blocks, switches with specializations, and interface modules. In addition, the organization was working on a continuous improvement approach and was involved in research and development activities. The study first identified twenty-six enablers of the GLSS strategy through an extensive literature review, and experts' opinions were subsequently used to determine the relevant enablers in the electrical component manufacturing industry. The selected enablers were further evaluated using the fuzzy decision-making trial and evaluation laboratory (DEMATEL) technique using experts, including technical experts, senior-level management experts, and one environmental consultant with different expertise in human resource management, technical aspects, and environmental management. Drawing on the fuzzy DEMATEL technique, the study identified environmental management systems (EMS), top management support, the organization's readiness to implement GLSS, market increase, organizational culture, and change management as the key factors in GLSS implementation in the discrete manufacturing industry (electrical products manufacturing firm).

Another study conducted in the cookware manufacturing industry in India developed, validated, and verified a GLSS framework (Tiwari et al. 2020). Using prior studies, the authors initially developed a framework that included lean, Six Sigma, and sustainability. As such, the framework comprised five stages and fifteen steps. Stage 1 involved planning and preparation, stage 2 included current evaluation, stage 3 comprised concept selection and implementation, stage 4 focused on implementation, evaluation, and knowledge sharing, and stage 5 entailed sustaining improvement. These stages require top management commitment; cross-functional teams; GLSS awareness; specific, measurable, achievable, relevant, and time-bound (SMART) objectives; a project charter; identifying improvements; data collection and analysis; and process monitoring and controlling for effective deployment of the framework. After developing the framework, the experts' opinions were considered using the Delphi technique to validate the framework. LSS experts who had expertise in the LSS black and green belts participated in the research study. The authors revised and modified the framework based on the experts' insights. Eleven experts from top management and managerial positions participated in the study. The proposed framework was shared with the experts and subsequently revised in two rounds of reviews. In the verification stage, the authors conducted a case study in a cookware manufacturing company. As such, the framework was implemented in one of the largest cookware manufacturing companies for three months. The company had affiliations in different regions, including the USA, Asia, and Europe, and it was ISO 9001 certified, a crucial factor in implementing LSS practices. The authors selected the company's engineering department and implemented the proposed framework in the photo paint production process, with new technologies installed to paint the products.

In this process, the metal sheets in the form of discs were transferred to the photo paint production lines from the main production line, where these discs were painted. Then, the painted discs were sent to the production line to manufacture different types of pans such as saucepans, skillets, and stockpots. The overall goal of the project was to minimize health and safety risks, adverse environmental effects, and operational costs. Along with the above objectives, the project aimed to address issues such as rejection rate, long setup times, unplanned stoppages, and defects in the painted products.

Another study adopted a sustainable-VSM methodology to investigate GLSS application (Faulkner and Badurdeen 2014). The study first developed a Sus-VSM model comprising environmental and social sustainability metrics. In this context, the authors conducted a comprehensive literature review to identify the relevant environmental and societal aspects of a discrete industry's manufacturing process. The environmental sustainability metrics included the process water consumption, raw material usage, and energy consumption. On the other hand, the social sustainability metrics comprised the physical and work environment information. After developing the framework, the authors validated it through a case study conducted in a company manufacturing satellite television dishes. The authors observed value-added and non-value-added activities in the manufacturing process, starting from the arrival of steel coils in the company, leading to the production of the dishes, and then the washing process to remove the oil and impurities, followed by the drying process. Finally, the painting was done, followed by the curing process.

The study revealed that the value-added time was around 32 min compared to the total lead time of over 12 days, highlighting the considerable non-value-added activities in the manufacturing process. Moreover, the results indicated high energy consumption (60%) in the washing process alone. In addition, the 64 gallons of water wasted per dish indicated a significant loss in natural resource consumption, leading to environmental sustainability issues. Moreover, the study results highlighted 35% of material loss in the form of scrap generated from the process. Overall, the Sus-VSM methodology helped the case company to visualize the value-added and non-value-added activities and understand the sustainability issues in the production line.

Another study utilized the GLSS strategy in the discrete manufacturing industry in India (Mishra et al. 2019). The study integrated VSM with simulation modelling to investigate the current state of the manufacturing process of bonnets and proposed an optimized process using ARENA software. The authors followed a case study approach conducted in the bonnet manufacturing industry in Bangalore, India. The company faced operational sustainability issues regarding cost reduction and on-time delivery; hence, it was keen to implement lean and sustainable initiatives. A team of experts and managers from different departments and the company's senior manager participated in the study. In addition, the top management also included LSS experts with prior experience in implementing LSS practices in the manufacturing industry. The team organized various brainstorming sessions to develop the current state of VSM.

Further, the study utilized the VSM methodology in conjunction with the simulation approach in the ARENA software. Moreover, the recorded values and observations of the process were used to develop the current state VSM. The rationale behind using the ARENA software was that it allows the "what-if analysis" and modifications in the process and product parameters to determine an optimal solution. In addition, the study used the Ishikawa diagram, FMEA, and Poka-yoke. Furthermore, a cost–benefit analysis was conducted to highlight the financial benefits of the future state VSM model.

The primary purpose of using VSM was to eliminate the bottlenecks and identify the opportunities in the process. The study highlighted the difference between the current and future states of the manufacturing process. The findings revealed a significant decrease in the cycle time of 30% using the proposed simulation model. Further, the study reported a reduction in the average number of work-in-progress pieces by ten. Along with improving the operational performance, the findings revealed a significant decline in the carbon footprint (83.7%) of the process, thus improving the environmental performance. Similarly, Ben Ruben et al. (2017) conducted a case study in the automotive industry to evaluate the effectiveness of lean Six Sigma from the environmental performance perspective, such as acidification, eutrophication, and ecotoxicity.

4.2.1.2 Cases of Green-Lean-Six Sigma Application in Process Industries

A process industry is classified as an industrial sector where the raw material is converted into finished products through successive continuous operations. Since plants continuously run to transform the raw materials into the output, any interruptions in the process can have far-reaching effects and affect product quality, leading to operational and environmental sustainability issues (Farrukh 2022; Taylor 2011). Different manufacturing operations in the continuous process industry include forming, mixing, and separating. The outputs are also in the form of continuous products such as gases, chemicals, pulp, rubber, and powders (Panwar et al. 2015; Prashar 2020; Yousefi and Rodriguez Hernandez 2019). Various industries working in the continuous process industry include chemical, rubber, petroleum, food, plastic, textile, cement, pharmaceutical, and beverage (Bening et al. 2021; Coppola et al. 2023; Dayaratne and Gunawardana 2015; Nguyen and Hens 2015; Singh et al. 2019a, b; Ugduler et al. 2021). Due to the nature of the manufacturing operations, use of hazardous raw materials, and excessive use of energy, water, and chemicals, this sector is facing significant environmental and social sustainability issues (Agrawal et al. 2024; Belamkar 2016; Chen et al. 2018; Duflou et al. 2012; Huisingh et al. 2015; Ncube et al. 2021). In addition, the execution of operational excellence strategies such as lean, Six Sigma, and total quality management is complex due to the characteristics of this sector, including the far-reaching errors, visibility issues, and long processing times (Farrukh 2022; Panwar et al. 2017).

While considering the sustainability issues of the firms operating in the process industry, prior studies have highlighted challenges such as excessive energy consumption, wastewater generation, and chemical consumption (Chen et al. 2023; Kharat et al. 2025; Omoloso et al. 2021; Suhandi and Chen 2023; Turker and Altuntas 2014). For example, the textile industry consumes large quantities of water and energy, releases air emissions, and discharges wastewater and toxic chemicals (Luoma et al. 2022; Reike et al. 2022; Stenton et al. 2021; Wren, 2022). Similarly, the cement industry releases GHG emissions such as $C0_2$ (Chotia et al. 2023; Gao et al. 2015; Ishak & Hashim, 2015; Seth et al. 2016). Conversely, the dairy industry generates significant emissions, particularly in milk production and processing (Bai et al. 2018; Chen et al. 2018; Lorenz et al. 2019; Salzman et al. 2017). In addition, the pharmaceutical industry is also facing significant sustainability issues, such as drug waste (Belal et al. 2024; Gerrans et al. 2023; Luo et al. 2021; McGlinchey et al. 2025; Suhandi and Chen 2023). A recent study highlighted that the pharmaceutical industry produces 52 million tons of CO_2 (Weaver et al. 2022). In addition, this industry uses large volumes of toxic chemicals and solvents to formulate medicines (Ashfaq et al. 2019; Hessel et al. 2022; Yaseen et al. 2021). According to an estimate, solvents comprise 80–90% of all materials used in drug manufacturing (Weaver et al. 2022). Despite this sector's significant adverse environmental impact, there is a lack of research investigating the deployment of operational excellence techniques with a combination of green practices (Farrukh et al. 2022; Panwar et al. 2018; Sharma et al. 2015). The following sections highlight the applications of GLSS strategy in the process industry.

A study conducted in the process industry in a developing economy presented a GLSS model using the insights obtained from the senior corporate managers working in flexible packaging companies (Farrukh et al. 2022). The study adopted a multiple case study approach and selected two large flexible packaging companies in Pakistan. The companies were manufacturing single and multi-layer packaging films in a combination of aluminium foil, paper, and plastic for manufacturing different customers' products such as tea, food, soap, detergents, pharmaceuticals, pesticides, personal care, and home care. The study classified the outcomes of implementing the GLSS strategy into three categories: (1) waste management and resource conservation, (2) emission control, and (3) environmental safety and long-term sustainability. In the waste management and resource-saving category, various outcomes included process and environmental waste reduction, efficient use of resources, less landfill, and reusing and recycling. Meanwhile, emission control comprises outcomes such as GHG and VOC emissions reduction. Finally, the environmental safety and long-term sustainability included environmental decision making, use of sustainable resources, environmental protection, and occupational health and safety. Overall, the study developed a holistic GLSS model that includes enablers to address environmental sustainability issues in the flexible packaging industry in Pakistan.

Another study, conducted in the dairy industry in China, performed a water footprint assessment based on ISO 14046 and highlighted that feed crop cultivation and cow breeding are the leading causes of water consumption, as well as dairy processing contributing significantly to water pollution. The study suggested the installation of wastewater recycling and treatment facilities and following stringent water quality standards as essential practices to reduce the water footprint in the dairy sector (Bai et al. 2018). Similarly, another study examined the application of LSS in the Norwegian dairy industry through a longitudinal study. The study revealed the benefits of LSS practices in reducing process waste, operational cost, wastewater, and defects, as well as increasing productivity and quality (Powell et al. 2017). Similarly, another study conducted in the ready-made garment industry in Bangladesh investigated the application of various GLSS tools, including the 5S, TPM, responsibility assignment matrix (RAM), and single minute exchange of die (SMED) (Mohammad et al. 2024). The authors conducted action research in three ready-made garment factories and observed the benefits of the above tools in reducing the changeover time, increasing productivity, and minimizing the lead time.

One study examined GLSS application in a plastic injection moulding factory in Iran producing plastic containers (Ershadi et al. 2021). The study highlighted the quality issues in the manufacturing operations related to the moulds, processing, devices and materials, which could lead to an increase in cost and material waste, as well as a reduction in the company's market share. The study first collected the relevant data through a comprehensive literature review and survey research, reading about GLSS projects. Moreover, the organization's documents were assessed to collect the data regarding machine performance. The selected company was one of the pioneering organizations in blowing and injecting polymers. The company's primary goals were to implement the GLSS project in the organization to improve productivity and simultaneously enhance its environmental and social performance objectives. The overall aim of the study was to develop a framework for evaluating and selecting appropriate GLSS projects in the organization. Based on the assessment of 28 indicators of GLSS project success, the committee selected two projects addressing the organization's objective of improving quality and sustainability while minimizing resource consumption, including energy, materials, and human efforts.

Similarly, a study by Tsarouhas and Sidiropoulou (2024) utilized a case study methodology in the olive packaging industry. The study used the DMAIC methodology to reduce the variation in the olives' drained weight in the glass jar with a capacity of 370 ml. Various tools such as Pareto diagram, SIPOCV, fishbone diagram, SPC (such as X bar and R charts), and process capability analysis were used in different phases of the DMAIC methodology to improve the quality of the products, reduce process variations, and increase the productivity which further resulted in waste reduction and conservation of materials such as chemicals, labels, olives, and caps (Tsarouhas and Sidiropoulou 2024).

The recent literature review also revealed the application of GLSS in the construction sector. For example, a recent study investigated the application of lean principles and combining them with building information modelling—a technological tool to reduce losses, enhance value, and streamline the flow of construction processes in the construction industry of Jordan (Aburumman et al. 2024). The study emphasized the potential benefits of increasing construction project efficiency, improving productivity, reducing waste, and minimizing costs. Similarly, another study examined the application of the GLSS strategy in the construction sector in the USA. The authors developed an integrated framework comprising various tools such as VSM, LCA, cause-effect diagram, and Pareto diagram to analyse the construction waste (Banawi and Bilec 2014).

4.2.2 Green-Lean-Six Sigma in Service Organizations

While GLSS is gaining traction in the service industry, the existing literature highlights a few studies on implementing GLSS in service organizations (Hariyani et al. 2025; Hundal et al. 2021; Jin et al. 2024; Rathi et al. 2023; Sijm-Eeken et al. 2024). For example, a study conducted in the retail industry in Italy examined GLSS implementation, as this sector (retail) contributes a significant quantity of waste (Marrucci et al. 2020). Since supermarkets can be essential in reducing waste and increasing resource efficiency, the retail sector has started implementing circular economy practices. The authors conducted a case study using an action research methodology to evaluate the waste management system in the Unicoop Firenze supermarket. The duration of the action research was nine months, and the participants were researchers, store managers, the head of sustainability, store department heads, and workers. In particular, the study followed the DMAIC methodology, integrated with the carbon footprint and cost benefits analysis to examine the environmental performance of different waste treatment methodologies, such as landfill and recycling. Based on the analysis, the authors suggested a suitable waste management system for the supermarket. As such, the study observed the benefits of using a GLSS strategy related to waste management, recycling, and circular economy, such as reducing the GHG emissions, including CO_2 (carbon dioxide), methane (CH4), and nitrous oxide (N_2O), and an increase in the volume of recycled waste.

4.2.2.1 Cases of Green-Lean-Six Sigma Application in the Healthcare Sector

The healthcare industry is generally considered an environmentally damaging or 'dirty industry' due to hazardous waste generation, but has started promoting green operations practices (Alshemari et al. 2020; Sathiya et al. 2023; Zhu et al. 2018). The literature review highlighted that this sector is also implementing operational excellence and management approaches such as lean manufacturing, Six Sigma, and total quality management to achieve operational objectives such as cost reduction, lead time reduction, and high quality (Dickson et al. 2009; Vasileiou et al. 2024; Zhu et al. 2018). Similarly, prior studies have highlighted the application of LSS in hospital emergency departments with key benefits including cost reduction, decrease in patients' waiting time, and increase in patient flow, increase in staff satisfaction, and improved cooperation and coordination between different functions (Costa et al. 2024; Dickson et al. 2009; Furukawa et al. 2016). Along with maintaining operational performance, this sector is trying to achieve environmental sustainability. However, the extant literature revealed limited studies investigating GLSS applications in the healthcare sector regarding greening healthcare operations.

For example, a study conducted in the US healthcare industry investigated the application of the GLSS strategy (Zhu et al. 2018). The authors first developed an adapted framework for integrating green and lean processes in the hospital supply chain. The framework was based on the traditional IDEF meta model comprising input, control, mechanism, and output. In addition, it included dimensions such as the current supply chain system, hospital value proposition, internal management, organizational structure, GLSS tools, and stakeholder collaboration. Then, three case studies were conducted in a hospital supply chain to uncover the green dimensions in the health care sector. These case studies were conducted in hospitals' linen operations, medication delivery systems, and sterile processing. The first case study in the linen operations in a hospital addressed outcomes such as more accurate demand and increased interaction among parties, such as EMS and nurses, in the new distribution system, which leads to better control and efficient use of resources such as linen, waste reduction, energy savings, and reduced carbon emissions from linen delivery trips. The second case study revealed reduced paper consumption, a decrease in missing medications, and minimal medication delivery trips, reducing paper waste, environmental waste, and carbon emissions in the medication delivery system. The third case addressed the potential benefits, such as reducing inventory waste, including missing and obsolete instruments in the sterile processing department, which leads to environmental and human safety benefits.

Similarly, another study conducted on the medication process in the hospital sector in Brazil revealed the significance of the GLSS strategy (Furukawa et al. 2016). The study adopted a longitudinal approach to investigate the environmental sustainability aspects in the medication process, such as excessive use of paper, labels, packaging, and medications. The study results demonstrate environmental sustainability improvement after implementing GLSS. The study reported a 74.8% reduction of chemical, sharps, and infectious waste, an 20% increase in non-recyclable waste, and a 33.3% increase in common recyclable waste in the pharmacy service. On the other hand, the nursing service achieved a 22.5% reduction of chemical, sharps, and infectious waste, an20% increase in common non-recyclable waste, and a 22.9% increase of common recyclable waste in the medical-surgical clinical unit. Overall, the study results revealed plastic bag reduction (37.1%), a reduction in leftover anti-inflammatory drugs (67.1%), an increase in removal of labels from plastic bags to be recycled (146.9%), and an increase in proper waste disposal (32.2%).

4.2.2.2 Cases of Green-Lean-Six Sigma Application in the Education Sector

While applying operational excellence techniques can improve sustainable performance in the service sector, the existing literature revealed a lack of application of GLSS in this domain (Lu et al. 2017; O'Reilly et al. 2019). In this context, a recent study recommended the application of GLSS in the higher education industry (Yadav et al. 2023). The authors highlighted that although the lean philosophy has been implemented successfully in the higher education sector, incorporating Six Sigma tools to investigate the variations in university business operations has gained less attention. However, using integrated LSS methodology can benefit the higher education industry in achieving business excellence objectives. The authors suggested that DMAIC is a useful methodology for addressing issues in the education sector and academic processes and removing the fundamental causes of the lack of business performance. Prior studies also suggested that GLSS application in the higher education sector can achieve better operational performance and address environmental sustainability in campus management operations (Klein et al. 2024). Nevertheless, limited research has been conducted in this sector (Laux et al. 2017; Lu et al. 2017; O'Reilly et al. 2019; Svensson et al. 2015).

Similarly, another study by O'Reilly et al. (2019) examined a GLSS application in a specific department in the higher education sector for reducing waste, minimizing cost, and improving operational efficiency. On the other hand, a study by Lu et al. (2017) found that the inclusion of the GLSS strategy as a holistic approach in curriculum development can enable students to understand sustainable consumption and production aspects, helping them become environmentally responsible citizens, and providing job opportunities in the dynamic job market. Similarly, a study conducted by Cudney et al. (2020) highlighted that GLSS implementation can help the higher education sector streamline its processes, reduce costs, and enhance sustainability. In addition, one study conducted in the higher education sector in Portugal utilized a quantitative research methodology to examine the relationship between lean practices and team performance in higher education management. The findings revealed the positive impact of continuous improvement on waste elimination. As such, the study showed that the employees' efforts to understand the processes, improve daily work, learn new ways of problem solving, and work standardization positively influence waste reduction in the higher education sector (Klein et al. 2024).

4.2.2.3 Application of Green-Lean-Six Sigma in SMEs

SMEs are recognized as key contributors to the economic development of any country (M. Singh et al. 2019a, b). This sector produces a variety of products and services to meet the demands of the local and international customers (Singh et al. 2019a, b). Along with economic growth, SMEs are recognized as a significant sector that creates jobs and innovates. According to the European Union, this sector has contributed to "85% of new employment in the past half-decade and accounted for 99% of business" (Mohan et al. 2023, p. 2). The categorization of SMEs varies from country to country (Ali et al. 2020). For example, organizations with fewer than 100 employees are considered SMEs in developed countries such as the USA. Alternatively, the EU Commission considers organizations with fewer than 250 employees to be SMEs. Conversely, in developing countries such as Pakistan, firms with fewer than 50 employees are considered SMEs (Ali et al. 2020).

On the other hand, poor performance and health of SMEs affect their country's economic growth. Due to its significance, this sector requires scholarly attention and particular focus from practitioners, policy makers, and decision makers (Ali et al. 2020). The importance of SMEs in a developing economy such as Pakistan can be understood from the facts indicating a total of 3.2 million SMEs in different cities of Pakistan, with a total contribution of approximately 30% to the country's GDP, 30% to overall exports, and 78% to industrial employment (Shah et al. 2018). Similarly, SMEs in India contribute around 40% of the overall exports and 45% of the manufacturing output in the economy (Mohan et al. 2023).

The review of literature revealed that a few studies have investigated the GLSS approach in SMEs (Ali et al. 2020; Mohan et al. 2025; Siegel et al. 2019; Sodhi et al. 2020; Thanki and Thakkar 2018). For example, a study conducted in the MSME industry in India investigated GLSS application (Singh et al. 2021). The study first identified 30 GLSS enablers through an extensive literature review and then categorized these according to the fundamental characteristics using experts' opinions. The experts were selected based on a relevant industrial and academic background. Building on the valuable insights from the experts, twenty-two enablers were finalized and synthesized into categories such as strategic, environmental, resources, cultural, and linkage enablers. The study highlighted significant enablers such as supportive organizational infrastructure, top management commitment, the organization's initiatives, employees' rewards and incentives, emission reduction initiatives, project leadership, and GLSS environmental awareness programmes and training.

Another study conducted in the SME sector in Pakistan examined the impact of GLSS on firm performance from the perspective of business, operational, and environmental performance (Ali et al. 2020). The study followed a quantitative methodology and conducted surveys in different regions of Pakistan. The data were collected from various industries, including automotive, service, manufacturing, and retail. The findings revealed that GLSS, comprising three operational excellence strategies, positively affects the organizations' environmental performance (Ali et al. 2020; Shah et al. 2018). Building on the literature review, the study used performance measures and designed a questionnaire relevant to the SMEs. The study utilized a quantitative research methodology using a survey for data collection since it aimed to investigate the GLSS application in the SME sector, including numerous businesses. In addition, the survey was conducted in the major cities of Punjab, including Lahore, Sialkot, and Faisalabad, due to the many SMEs operating in this region (Zafar and Mustafa 2017).

Overall, the findings revealed the positive impact of green, lean, and Six Sigma strategies on the environmental performance of SMEs; nevertheless, the study showed no effect on the SMEs' operational and business performance from using these strategies. In addition, the authors explained the lack of execution of JIT practices in the SME sector of Pakistan as due to the lack of awareness of the JIT concept in the large manufacturing companies in Pakistan that place orders for their supplies and raw materials with the SMEs. Though 78% of SMEs are aware of the JIT technique, the implementation rate is only around 2% (Ali et al. 2020). However, training and awareness sessions on the significance of lean practices can improve organizations' business and operational performance.

Another case study was conducted in an SME wood product manufacturing facility in the US, producing cutting boards using the batch production system with a batch size of 6 cutting boards (Tasdemir and Gazo 2019). The authors validated the sustainability benchmarking tool (SBT) framework to achieve sustainability performance by optimally utilizing natural resources such as wood to manufacture value-added products. The study reported findings on sustainability performance from pre- and post-improvement phases to validate the proposed SBT framework. As such, the authors followed a six-step implementation approach developed in their prior study (Tasdemir et al. 2019). In step 1, the stakeholders, their expectations, and the cost of non-conformity and compliance were identified. Step 2 identified the organizational problems, and the supply chain process was developed to understand the company's operations. In this stage, a team of three participants was formed, and they arranged weekly meetings and brainstorming sessions to identify and record the issues in the manufacturing operations. The study used SIPOC and process mapping in the first two stages.

Next, the relevant historical data were obtained and evaluated in step 3, leading to the development of the current state VSM. In step 4, the environmental risks were identified and ranked according to their severity and occurrence using a risk probability/impact matrix. Moreover, the risk control mechanisms were identified accordingly. In this step, the experts with the Six Sigma black belt certification and ISO 9001:2008 internal audit conducted the Gemba walks. In addition, a measurement system analysis (MSA) was performed, including the overall defect rate, product dimensions, cycle time, waste measurement (such as glue, wood, mineral oil, and kraft paper), and noise level to ensure the accuracy of the collected data and address the repeatability and reproducibility concerns in the study.

Later in step 4, scores were given to 18 KPIs to identify the strengths and weaknesses of the manufacturing system, e.g. scores less than seven were marked as weaknesses, and scores greater than nine were considered opportunities. The fishbone diagram, FMEA, regression analysis, and 5 whys were used to identify the key reasons behind the weaknesses in step 5. In addition, the brainstorming sessions using the impact/effort matrix were conducted to discuss and document the improvement ideas. Further, the control mechanisms relevant to the improvement ideas were developed to ensure standardization. Finally, in stage 6, the comparison between the pre-improvement and post-improvement stages was conducted, and the findings presented. Overall, the study highlighted sustainability benefits in the form of a 58% decrease in transportation and motion distance, a net reduction in CO_2 emissions and energy footprint impact, and a 72.03% reduction in total solid waste generation.

Similarly, another study investigated GLSS application in SMEs and developed an integrated GLSS framework comprising eight steps (Yadav and Gahlot 2022). The first step emphasized selecting a GLSS project focused on reducing environmental emissions, defects, and rejection, and integrating customers' voices into projects. In addition, the scope, objectives, and overall aim of the GLSS project are determined in this stage. This step is crucial in the GLSS deployment in SMEs, as the project scope should be adequately defined and aligned with the sustainability targets.

On the other hand, step 2 emphasizes the problem description as outlining the relevant problem helps filter the insignificant and least important issues. In addition, the problem statement should focus on defect reduction, air and waste reduction or mitigation, and producing quality and environmentally friendly products at reduced cost. In this step, the organizations can use several tools such as mind maps, flowcharts, concept maps, idea maps, strategy maps, concept maps, etc. These maps help determine "what happens at each stage of a process and how each event affects it" (Yadav and Gahlot 2022, p. 10).

Step 3 included the data collection process on the current state of the project while focusing on the critical performance metrics related to the quality and environmental aspects. The data collection further led to the identification of the root causes of the defects. Various GLSS tools, including process capability analysis, SPC, EVSM, Pareto chart, and LCA, can be used to evaluate CO_2 emissions and raw material and energy consumption. Step 4 focused on mapping the current process to improve the efficiency of manufacturing operations in an organization. In this regard, process maps assist in brainstorming ideas, enhancing communication among team members, and documenting the process activities. In addition, various elements, including activity steps, functions, people, input/output, decision points, and time in the process maps, help identify the process bottlenecks and delays. This step identified SIPOC and VSM as significant tools for developing process maps.

Step 5 identified the key reasons for process inefficiencies leading to defects, rejections, and environmental emissions in the GLSS projects. Both the non-value-added and value-added activities from both the customer and business perspectives are identified in this step. Various GLSS tools, such as cause-and-effect diagrams, LCA, 5 whys, failure mode and effects analysis (FMEA), Pareto charts, regression analysis, and principal component analysis, are used to determine the root causes of sustainability issues in GLSS projects. Step 6 emphasized the development of solutions and the implementation of suitable ones. In addition, appropriate employee training is also ensured during this step to effectively deploy the selected solution. Various tools, including the SPC, 5S, cause-and-effect diagram, DOE, LCA, Pugh matrix, and Kaizen, can be used in this regard. Step 7 is directed towards controlling the process to ensure the continuous implementation of the selected solution and preserving the positive impact. Moreover, the results are shared with the employees. Finally, step 8 focused on sustaining the implemented solution in the process. Various tools, including the SOPs, TPM, FMEA, visual management, and Poka-yoke, were used in this step. Moreover, VSM and LCA can be applied to re-examine the process after deploying the solution to identify the improvements.

4.3 Summary

This chapter demonstrated the application of GLSS in manufacturing and service organizations. Specifically, it discussed cases, highlighting discrete and process manufacturing industries using GLSS applications to achieve operational and environmental performance. Further, cases in the service sector, including the healthcare and higher education sectors, were also presented to illustrate how GLSS is becoming an emerging paradigm in these contexts. Finally, this chapter also shed light on the role of GLSS in Small and Medium-sized Enterprises (SMEs).

References

Aburumman, M. O., Sweis, R., & Sweis, G. J. (2024). Investigating building information modelling (BIM) and lean construction: the potential BIM-lean interactions synergy and integration in the Jordanian construction industry. *International Journal of Lean Six Sigma*, *15*(2), 400-438. https://doi.org/10.1108/IJLSS-01-2023-0013

Agrawal, M., Bansal, A., Khandelwal, V., & Bansal, N. (2024). Sustainable pharma: The need, current status and mission for the future. *Scripta Medica*, *55*(4), 489-499. https://doi.org/10.5937/scriptamed55-51612

Ali, Y., Younus, A., Khan, A. U., & Pervez, H. (2020a). Impact of Lean, Six Sigma and environmental sustainability on the performance of SMEs. *International Journal of Productivity and Performance Management*, *70*(8), 2294-2318. https://doi.org/10.1108/ijppm-11-2019-0528

Alshemari, A., Breen, L., Quinn, G., & Sivarajah, U. (2020). Can we create a circular pharmaceutical supply chain (CPSC) to reduce medicines waste? *Pharmacy*, *8*(4).

Ashfaq, M., Li, Y., Rehman, M. S. U., Zubair, M., Mustafa, G., Nazar, M. F.,…Sun, Q. (2019). Occurrence, spatial variation and risk assessment of pharmaceuticals and personal care products in urban wastewater, canal surface water, and their sediments: a case study of Lahore, Pakistan. *Sci Total Environ*, *688*, 653–663. https://doi.org/10.1016/j.scitotenv.2019.06.285

Bai, X., Ren, X., Khanna, N. Z., Zhou, N., & Hu, M. (2018). Comprehensive water footprint assessment of the dairy industry chain based on ISO 14046: A case study in China. *Resour. Conserv. Recycl.*, *132*, 369-375. https://doi.org/10.1016/j.resconrec.2017.07.021

Banawi, A., & Bilec, M. M. (2014). A framework to improve construction processes: Integrating Lean, Green and Six Sigma. *International Journal of Construction Management*, *14*(1), 45-55. https://doi.org/10.1080/15623599.2013.875266

Belal, Md M., Shukla, V., & Balasubramanian, S. (2024). Greening the Pharmaceutical Supply Chain. *Business Strategy and the Environment*, *34*(2), 1917-1948. https://doi.org/10.1002/bse.4077

Belamkar, N. (2016). development of green high performance liquid chromatographymethod for determination for ranitidine hydrochloride in solid dosage form by applying lean sigma principles. *Research Journal of Pharmaceutical, Biological and Chemical Sciences.*

Ben Ruben, R., Vinodh, S., & Asokan, P. (2017). Implementation of Lean Six Sigma framework with environmental considerations in an Indian automotive component manufacturing firm: a case study. *Production Planning & Control, 28*(15), 1193–1211. https://doi.org/10.1080/09537287.2017.1357215

Bening, C. R., Pruess, J. T., & Blum, N. U. (2021). Towards a circular plastics economy: Interacting barriers and contested solutions for flexible packaging recycling. *Journal of Cleaner Production, 302.* https://doi.org/10.1016/j.jclepro.2021.126966

Bhamu, J., Khandelwal, A., & Sangwan, K. S. (2013). Lean manufacturing implementation in an automated production line: a case study. *International Journal of Services and Operations Management, 15*(4), 411-429.

Chen, G. Q., Talebi, S., Gras, S. L., Weeks, M., & Kentish, S. E. (2018). A review of salty waste stream management in the Australian dairy industry. *J Environ Manage, 224*, 406-413. https://doi.org/10.1016/j.jenvman.2018.07.056

Chen, X., Xu, L., Ren, Z., Jia, F., & Yu, Y. (2023). Sustainable supply chain management in the leather industry: a systematic literature review. *International Journal of Logistics Research and Applications, 26*(12), 1663-1703.

Chiarini, A. (2014). Sustainable manufacturing-greening processes using specific Lean Production tools: an empirical observation from European motorcycle component manufacturers. *Journal of Cleaner Production, 85*, 226-233. https://doi.org/10.1016/j.jclepro.2014.07.080

Chotia, V., Soni, S., Jain, G., & Papa, A. (2023). Barriers for adoption of green supply chain management in cement industry: an interpretive structural modelling (ISM) approach. *Annals of Operations Research*, 1–34.

Coppola, C., Vollero, A., & Siano, A. (2023). Developing dynamic capabilities for the circular economy in the textile and clothing industry in Italy: A natural-resource-based view. *Business Strategy and the Environment, 32*(7), 4798-4820.

Costa, F., Portioli-Staudacher, A., Alemsan, N., & Tortorella, G. L. (2024). Readiness level assessment for lean Six Sigma implementation in the healthcare sector. *International Journal of Lean Six Sigma, 15*(1), 131-152. https://doi.org/10.1108/IJLSS-02-2023-0031

Cristina De Stefano, M., Montes-Sancho, M. J., & Busch, T. (2016). A natural resource-based view of climate change: Innovation challenges in the automobile industry. *Journal of Cleaner Production, 139*, 1436-1448. https://doi.org/10.1016/j.jclepro.2016.08.023

Cudney, E. A., Venuthurumilli, S. S. J., Materla, T., & Antony, J. (2020). Systematic review of Lean and Six Sigma approaches in higher education. *Total Quality Management & Business Excellence, 31*(3–4), 231–244.

Dayaratne, S. P., & Gunawardana, K. D. (2015). Carbon footprint reduction: a critical study of rubber production in small and medium scale enterprises in Sri Lanka. *Journal of Cleaner Production, 103*, 87-103. https://doi.org/10.1016/j.jclepro.2014.09.101

Deepak, S., & Dharmendra, S. (2025). Green Lean Six Sigma Barriers and Efficacy in Ceramic Manufacturing Industries: A PLS-SEM Study. *Evergreen, 12*(1), 176-190. https://doi.org/10.5109/7342448

Dickson, E. W., Singh, S., Cheung, D. S., Wyatt, C. C., & Nugent, A. S. (2009). Application of lean manufacturing techniques in the Emergency Department. *J Emerg Med, 37*(2), 177-182. https://doi.org/10.1016/j.jemermed.2007.11.108

Duflou, J. R., Sutherland, J. W., Dornfeld, D., Herrmann, C., Jeswiet, J., Kara, S.,...Kellens, K. (2012). Towards energy and resource efficient manufacturing: a processes and systems approach. *CIRP Annals, 61*, 587–609.

Ershadi, M. J., Qhanadi Taghizadeh, O., & Hadji Molana, S. M. (2021). Selection and performance estimation of Green Lean Six Sigma Projects: a hybrid approach of technology readiness level, data envelopment analysis, and ANFIS. *Environmental Science and Pollution Research.* https://doi.org/10.1007/s11356-021-12595-5

Farrukh, A., Mathrani, S., & Sajjad, A. (2022). Managerial perspectives on green-lean-six sigma adoption in the flexible packaging industry: empirical evidence from an emerging economy. *Journal of Manufacturing Technology Management, ahead-of-print*(ahead-of-print). https://doi.org/10.1108/JMTM-02-2022-0080

Farrukh, A. (2022). A green-lean-six sigma model for environmental performance in manufacturing organizations: a study of a developed and developing nation: a thesis presented in partial fulfilment of the requirement for the degree of Doctor of Philosophy in Engineering Management, School of Food and Advanced Technology, Massey University, Auckland, New Zealand.

Faulkner, W., & Badurdeen, F. (2014). Sustainable Value Stream Mapping (Sus-VSM): methodology to visualize and assess manufacturing sustainability performance. *Journal of Cleaner Production, 85*, 8-18. https://doi.org/10.1016/j.jclepro.2014.05.042

Furukawa, P. D. O., Cunha, I. C. K. O., & Pedreira M., D. L. (2016). Evaluation of environmentally sustainable actions in the medication process. *Rev Bras Enferm, 69*(1), 16–22. https://doi.org/10.1590/0034-7167.2016690103i

Gao, T., Shen, L., Shen, M., Chen, F., Liu, L., & Gao, L. (2015). Analysis on differences of carbon dioxide emission from cement production and their major determinants. *Journal of Cleaner Production, 103*, 160-170.

Gerrans, J., Donyai, P., Finlay, K., & Sherratt, R. S. (2023). An efficient smart pharmaceutical packaging technology framework to assess the quality of returned medication through non-intrusively recording storage conditions after dispensation. *Technologies, 11*(3). https://doi.org/10.3390/technologies11030075

Gholami, H., Jamil, N., Mat Saman, M. Z., Streimikiene, D., Sharif, S., & Zakuan, N. (2020). The application of green lean six sigma. *Business Strategy and the Environment.*

Hariyani, D., Hariyani, P., Mishra, S., & Sharma, M. K. (2025). Drivers for the adoption of integrated sustainable green lean six sigma agile service system (ISGLSASS) in the service organizations. *Sustainable Futures, 9.* https://doi.org/10.1016/j.sftr.2025.100571

Hessel, V., Tran, N. N., Asrami, M. R., Tran, Q. D., Long, N. V. D., Escribà-Gelonch, M.,…Sundmacher, K. (2022). Sustainability of green solvents–review and perspective. *Green Chemistry, 24*(2), 410–437.

Huisingh, D., Zhang, Z., Moore, J. C., Qiao, Q., & Li, Q. (2015). Recent advances in carbon emissions reduction: policies, technologies, monitoring, assessment and modeling. *Journal of Cleaner Production, 103*, 1-12. https://doi.org/10.1016/j.jclepro.2015.04.098

Hundal, G. S., Thiyagarajan, S., Alduraibi, M., Laux, C. M., Furterer, S. L., Cudney, E. A., & Antony, J. (2021). Lean Six Sigma as an organizational resilience mechanism in health care during the era of COVID-19. *International Journal of Lean Six Sigma, ahead-of-print*(ahead-of-print). https://doi.org/10.1108/ijlss-11-2020-0204

Ishak, S. A., & Hashim, H. (2015). Low carbon measures for cement plant–a review. *Journal of Cleaner Production, 103*, 260-274.

Jin, H., Qu, P., Zhou, Y., Qu, Q., & Li, H. (2024). Green Lean Six Sigma practice in the industrial and service sectors: from the systematic literature review to an integration diagram. *Production Planning & Control*, 1-26. https://doi.org/10.1080/09537287.2024.2369779

Kharat, M. G., Kapoor, S., Parhi, S., Kharat, M. G., & Pandey, S. (2025). Operationalizing sustainability in pharmaceuticals: green supply chain metrics for circular economy. *Sustainable Futures, 9.* https://doi.org/10.1016/j.sftr.2024.100413

Klein, L. L., Moyano-Fuentes, J., Vieira, K. M., & Marçal, D. R. (2024). An exploratory study of the relationships between Lean practices and team performance in higher education. *International Journal of Lean Six Sigma, 15*(2), 372-399. https://doi.org/10.1108/IJLSS-08-2022-0168

Laux, C., Li, N., Seliger, C., & Springer, J. (2017). Impacting Big Data analytics in higher education through Six Sigma techniques. *International Journal of Productivity and Performance Management, 66*(5), 662-679. https://doi.org/10.1108/ijppm-09-2016-0194

Liu, F., & Miao, Z. (2006). The application of RFID technology in production control in the discrete manufacturing industry. 2006 IEEE International Conference on Video and Signal Based Surveillance,

Lorenz, H., Reinsch, T., Hess, S., & Taube, F. (2019). Is low-input dairy farming more climate friendly? A meta-analysis of the carbon footprints of different production systems. *Journal of Cleaner Production, 211*, 161-170. https://doi.org/10.1016/j.jclepro.2018.11.113

Lu, J., Laux, C., & Antony, J. (2017). Lean Six Sigma leadership in higher education institutions. *International Journal of Productivity and Performance Management, 66*(5), 638-650.

Luo, Y., Reimers, K., Yang, L., & Lin, J. (2021). Household drug management practices of residents in a second-tier city in China: Opportunities for reducing drug waste and environmental pollution. *International Journal of Environmental Research and Public Health, 18*(16), 8544.

Luoma, P., Penttinen, E., Tapio, P., & Toppinen, A. (2022). Future images of data in circular economy for textiles. *Technological Forecasting and Social Change, 182.* https://doi.org/10.1016/j.techfore.2022.121859

Maldonado-Guzmán, G., Garza-Reyes, J. A., & Pinzón-Castro, Y. (2020). Eco-innovation and the circular economy in the automotive industry. *Benchmarking: An International Journal, 28*(2), 621–635. https://doi.org/10.1108/bij-06-2020-0317

Marrucci, L., Marchi, M., & Daddi, T. (2020). Improving the carbon footprint of food and packaging waste management in a supermarket of the Italian retail sector. *Waste Manag, 105*, 594-603. https://doi.org/10.1016/j.wasman.2020.03.002

McGlinchey, Y., Iqbal, J., Trubetskaya, A., Thenarasu, M., & McDermott, O. (2025). Design for green lean six sigma to improve sustainability in the pharmaceutical industry—a case study. *Production & Manufacturing Research, 13*(1). https://doi.org/10.1080/21693277.2025.2461010

Mishra, A. K., Sharma, A., Sachdeo, M., & K, J. (2019). Development of sustainable value stream mapping (SVSM) for unit part manufacturing. *International Journal of Lean Six Sigma, 11*(3), 493–514. https://doi.org/10.1108/ijlss-04-2018-0036

Mohammad, A., Hamja, A., & Hasle, P. (2024). Reduction of changeover time through SMED with RACI integration in garment factories. *International Journal of Lean Six Sigma, 15*(2), 201-219. https://doi.org/10.1108/IJLSS-10-2021-0176

Mohan, J., Kaswan, M. S., & Rathi, R. (2023). Identification and investigation into the barriers to Green Lean Six Sigma implementation: a micro small and medium enterprises perspective. *International Journal on Interactive Design and Manufacturing (IJIDeM).* https://doi.org/10.1007/s12008-023-01551-x

Mohan, J., Kaswan, M. S., & Rathi, R. (2025). An analysis of green lean six sigma deployment in MSMEs: a systematic literature review and conceptual implementation framework. *The TQM Journal, 37*(3), 747-777.

Ncube, L. K., Ude, A. U., Ogunmuyiwa, E. N., Zulkifli, R., & Beas, I. N. (2021). An overview of plastic waste generation and management in food packaging industries. *Recycling, 6*(1), 12.

Nguyen, Q. A., & Hens, L. (2015). Environmental performance of the cement industry in Vietnam: the influence of ISO 14001 certification. *Journal of Cleaner Production, 96*, 362-378. https://doi.org/10.1016/j.jclepro.2013.09.032

O'Reilly, S. J., Healy, J., Murphy, T., & Ó'Dubhghaill, R. (2019). Lean Six Sigma in higher education institutes: an Irish case study. *International Journal of Lean Six Sigma, 10*(4), 948-974.

Omoloso, O., Mortimer, K., Wise, W. R., & Jraisat, L. (2021). Sustainability research in the leather industry: A critical review of progress and opportunities for future research. *Journal of Cleaner Production, 285*, 125441.

Panwar, A., Nepal, B. P., Jain, R., & Rathore, A. P. S. (2015). On the adoption of lean manufacturing principles in process industries. *Production Planning & Control, 26*(7), 564-587. https://doi.org/10.1080/09537287.2014.936532

Panwar, A., Jain, R., Rathore, A. P. S., Nepal, B., & Lyons, A. C. (2017). The impact of lean practices on operational performance – an empirical investigation of Indian process industries. *Production Planning & Control, 29*(2), 158-169. https://doi.org/10.1080/09537287.2017.1397788

Panwar, A., Jain, R., Rathore, A. P. S., Nepal, B., & Lyons, A. (2018). The impact of lean practices on operational performance–an empirical investigation of Indian process industries. *Production Planning & Control, 29*(2), 158-169.

Parmar, P. S., & Desai, T. N. (2020). Evaluating Sustainable Lean Six Sigma enablers using fuzzy DEMATEL: a case of an Indian manufacturing organization. *Journal of Cleaner Production, 265.* https://doi.org/10.1016/j.jclepro.2020.121802

Powell, D., Lundeby, S., Chabada, L., & Dreyer, H. (2017). Lean Six Sigma and environmental sustainability: the case of a Norwegian dairy producer. *International Journal of Lean Six Sigma, 8*(1), 53-64. https://doi.org/10.1108/ijlss-06-2015-0024

Prasad, M. M., Dhiyaneswari, J. M., Jamaan, J. R., Mythreyan, S., & Sutharsan, S. M. (2020). A framework for lean manufacturing implementation in Indian textile industry. *Materials Today: Proceedings, 33,* 2986-2995.

Prashar, A. (2020). Adopting Six Sigma DMAIC for environmental considerations in process industry environment. *The TQM Journal, ahead-of-print*(ahead-of-print). https://doi.org/10.1108/tqm-09-2019-0226

Rathi, R., Kaswan, M. S., Antony, J., Cross, J., Garza-Reyes, J. A., & Furterer, S. L. (2023). Success factors for the adoption of green lean six sigma in healthcare facility: an ISM-MICMAC study. *International Journal of Lean Six Sigma, 14*(4), 864-897. https://doi.org/10.1108/IJLSS-02-2022-0042

Reike, D., Hekkert, M. P., & Negro, S. O. (2022). Understanding circular economy transitions: The case of circular textiles. *Business Strategy and the Environment, 32*(3), 1032-1058. https://doi.org/10.1002/bse.3114

Salzman, S., Scarborough, H., & Allinson, G. (2017). Exploring the economic viability of using constructed wetlands to manage waste-water in the dairy industry. *Australasian Journal of Environmental Management, 24*(3), 276-288. https://doi.org/10.1080/14486563.2017.1328711

Sasikumar, A., & Acharya, P. (2020). An operational approach to gauge the lean performance in bias tyre manufacturing: a case study. *International Journal of Productivity and Quality Management, 31*(3), 369-389.

Sathiya, V., Nagalakshmi, K., Jeevamalar, J., Anand Babu, R., Karthi, R., Acevedo-Duque, A.,…Ramabalan, S. (2023). Reshaping healthcare supply chain using chain-of-things technology and key lessons experienced from COVID-19 pandemic. *Socioecon Plann Sci, 85,* 101510. https://doi.org/10.1016/j.seps.2023.101510

Seth, D., Shrivastava, R., & Shrivastava, S. (2016). An empirical investigation of critical success factors and performance measures for green manufacturing in cement industry. *Journal of Manufacturing Technology Management, 27*(8), 1076-1101.

Shah, D., & Syed, A. J. I. P. C. o. P. (2018). Framework for SME sector development in Pakistan. *1*(1), 21–23.

Sharma, D., & Singh, D. (2025). Green lean six sigma barriers and efficacy in ceramic manufacturing industries: a PLS-SEM study.

Sharma, V. K., Anish Sachdeva, D. V. S. a. D., Chandana, p., & Bhardwaj, A. (2015). Critical factors analysis and its ranking for implementation of GSCM in Indian dairy industry. *Journal of Manufacturing Technology Management, 26*(6), 911–922. https://doi.org/10.1108/jmtm-03-2014-0023

Siegel, R., Antony, J., Garza-Reyes, J. A., Cherrafi, A., & Lameijer, B. (2019). Integrated green lean approach and sustainability for SMEs: from literature review to a conceptual framework. *Journal of Cleaner Production, 240.* https://doi.org/10.1016/j.jclepro.2019.118205

Sijm-Eeken, M., Greif, A., Peute, L., & Jaspers, M. (2024). Implementation of Green Lean Six Sigma in Dutch Healthcare: A Qualitative Study of Healthcare Professionals' Experiences. *Nursing Reports, 14*(4), 2877-2895. https://doi.org/10.3390/nursrep14040210

Singh, J., Sung, K., Cooper, T., West, K., & Mont, O. (2019a). Challenges and opportunities for scaling up upcycling businesses—the case of textile and wood upcycling businesses in the UK. *Resources, Conservation and Recycling, 150.* https://doi.org/10.1016/j.resconrec.2019.104439

Singh, M., Kumar, P., & Rathi, R. J. T. T. J. (2019). Modelling the barriers of Lean Six Sigma for Indian micro-small medium enterprises: an ISM and MICMAC approach. *31*(5), 673-695.

Singh, M., Rathi, R., & Garza-Reyes, J. A. (2021). Analysis and prioritization of Lean Six Sigma enablers with environmental facets using best worst method: a case of Indian MSMEs. *Journal of Cleaner Production, 279.* https://doi.org/10.1016/j.jclepro.2020.123592

Sodhi, H. S., Singh, D., & Singh, B. J. (2020). A conceptual examination of Lean, Six Sigma and Lean Six Sigma models for managing waste in manufacturing SMEs. *World Journal of Science, Technology and Sustainable Development, 17*(1), 20-32. https://doi.org/10.1108/wjstsd-10-2019-0073

Stenton, M., Houghton, J. A., Kapsali, V., & Blackburn, R. S. (2021). The potential for regenerated protein fibres within a circular economy: lessons from the past can inform sustainable innovation in the textiles industry. *Sustainability, 13*(4). https://doi.org/10.3390/su13042328

Suhandi, V., & Chen, P.-S. (2023). Closed-loop supply chain inventory model in the pharmaceutical industry toward a circular economy. *Journal of Cleaner Production, 383.* https://doi.org/10.1016/j.jclepro.2022.135474

Svensson, C., Jiju Antony, P., Antony, J., Ba-Essa, M., Bakhsh, M., & Albliwi, S. (2015). A Lean Six Sigma program in higher education. *International Journal of Quality & Reliability Management, 32*(9), 951-969. https://doi.org/10.1108/ijqrm-09-2014-0141

Tasdemir, C., Gazo, R., & Quesada, H. J. (2019). Sustainability benchmarking tool (SBT): theoretical and conceptual model proposition of a composite framework. *Environment, Development and Sustainability.* https://doi.org/10.1007/s10668-019-00512-3

Tasdemir, C., & Gazo, R. (2019). Validation of sustainability benchmarking tool in the context of value-added wood products manufacturing activities. *Sustainability, 11*(8). https://doi.org/10.3390/su11082361

Taylor, R. (2011). The economic impact of poor power quality to New Zealand.

Thanki, & Thakkar, J. (2018). Interdependence analysis of lean-green implementation challenges: a case of Indian SMEs. *Journal of Manufacturing Technology Management.* https://doi.org/10.1108/jmtm-04-2017-0067

Tiwari, P., Sadeghi, J. K., & Eseonu, C. (2020). A sustainable lean production framework with a case implementation: Practice-based view theory. *Journal of Cleaner Production, 277*, 123078.

Tsarouhas, P., & Sidiropoulou, N. (2024). Application of Six Sigma methodology using DMAIC approach for a packaging olives production system: a case study. *International Journal of Lean Six Sigma, 15*(2), 247-273. https://doi.org/10.1108/IJLSS-06-2022-0140

Turker, D., & Altuntas, C. (2014). Sustainable supply chain management in the fast fashion industry: An analysis of corporate reports. *European Management Journal, 32*(5), 837-849. https://doi.org/10.1016/j.emj.2014.02.001

Ugduler, S., De Somer, T., Van Geem, K. M., Roosen, M., Kulawig, A., Leineweber, R., & De Meester, S. (2021). Towards a Better Understanding of Delamination of Multilayer Flexible Packaging Films by Carboxylic Acids. *ChemSusChem.* https://doi.org/10.1002/cssc.202002877

Vasileiou, A., Sfakianaki, E., & Tsekouropoulos, G. (2024). Exploring sustainability and efficiency improvements in healthcare: a qualitative study. *Sustainability, 16*(19). https://doi.org/10.3390/su16198306

Vinnari, V. (2024). Operational economic and environmental reporting in mining industry information management system.

Weaver, E., O'Hagan, C., & Lamprou, D. A. (2022). The sustainability of emerging technologies for use in pharmaceutical manufacturing. *Expert Opin Drug Deliv, 19*(7), 861-872. https://doi.org/10.1080/17425247.2022.2093857

Wren, B. (2022). Sustainable supply chain management in the fast fashion industry: a comparative study of current efforts and best practices to address the climate crisis. *Cleaner Logistics and Supply Chain, 4.* https://doi.org/10.1016/j.clscn.2022.100032

Yadav, V., & Gahlot, P. (2022). Green Lean Six Sigma sustainability-oriented framework for small and medium enterprises. *International Journal of Quality & Reliability Management, 39*(7), 1787-1807.

Yadav, V., Gahlot, P., Rathi, R., Yadav, G., Kumar, A., & Kaswan, M. S. (2021). Integral measures and framework for green lean six sigma implementation in manufacturing environment. *International Journal of Sustainable Engineering, 14*(6), 1319-1331. https://doi.org/10.1080/19397038.2021.1970855

Yadav, V., Kumar, V., Gahlot, P., Mittal, A., Kaswan, M. S., Garza-Reyes, J. A.,...Owad, A. A. (2023). Exploration and mitigation of green lean six sigma barriers: a higher education institutions perspective. *The TQM Journal*. https://doi.org/10.1108/tqm-03-2023-0069

Yaseen, G., Ahmad, M., Zafar, M., Akram, A., Sultana, S., Kilic, O., & Sonmez, G. D. (2021). Current status of solvents used in the pharmaceutical industry. In *Green Sustainable Process for Chemical and Environmental Engineering and Science* (pp. 195–219). Elsevier.

Yousefi, A., & Rodriguez Hernandez, M. (2019). Using a system theory based method (STAMP) for hazard analysis in process industry. *Journal of Loss Prevention in the Process Industries, 61*, 305-324. https://doi.org/10.1016/j.jlp.2019.06.014

Zafar, A., & Mustafa, S. (2017). SMEs and its role in economic and socio-economic development of Pakistan. *International Journal of Academic Research in Accounting, Finance and Management Sciences, 6*(4).

Zhu, Q., Johnson, S., & Sarkis, J. (2018). Lean six sigma and environmental sustainability: a hospital perspective. *Supply Chain Forum: An International Journal, 19*(1), 25-41. https://doi.org/10.1080/16258312.2018.1426339

Chapter 5
Future Outlook and Emerging Trends in Green-Lean-Six Sigma

5.1 Introduction

This chapter examines the emerging trends affecting and transforming the adoption and application of Green-Lean-Six Sigma (GLSS) in industrial and service sector organizations. It discusses diverse trends shaping the GLSS phenomenon, and the most dominant forces among these include sustainability, the adoption of advanced technologies such as digitalization and Industry 4.0, the circular economy in the industrial sector to enhance operational excellence and sustainable business performance, and the interplay between GLSS and global supply chain operations and contemporary project management domains.

5.2 Integrating Green-Lean-Six Sigma with Emerging Trends

Over the past few years, the GLSS approach has received increasing recognition and acceptance in the manufacturing sector. However, as discussed in Chap. 4, the application of the GLSS concept has transcended beyond manufacturing and has been widely acknowledged by service sector organizations, including healthcare (Ahmed et al. 2024; Rathi et al. 2023), finance (Vashishth et al. 2024), and education (Yadav et al. 2024), which are increasingly incorporating GLSS approaches and frameworks in their business processes and operational activities (e.g., Madzík et al. 2025). The competitive business environment and serious environmental issues largely stemming from stakeholders' expectations continue to pressure both service and industrial sectors to embrace GLSS practices to enhance operational performance and balance dynamic and complex stakeholder requirements around sustainability and social responsibility (Jin et al. 2025; Siegel et al. 2019). However, Jin et al. (2025, p. 394) argued that "most GLSS frameworks are primarily designed for the industrial

A. Farrukh and A. Sajjad, *A Manager's Guide to Green-Lean-Six Sigma Application*,
https://doi.org/10.1007/978-3-032-21111-8_5

sector, and service organizations have yet to systematically grasp the application of GLSS tools".

Jin et al. (2025) highlighted diverse applications of GLSS in the service and industrial sectors. A wide range of approaches, practices, and tools is becoming more pronounced and increasingly popular, influencing the development of GLSS research and real-world practice. In a recent review, Jin et al. (2025) identified 51 salient GLSS practices and tools, including both qualitative and quantitative approaches used by service and industrial organizations to enhance their operational and sustainability performance. The authors noted that the industrial sector has employed more tools (n = 43) than the service sector (n = 34). Further, some tools are more frequently employed in the industrial sector than in the service sector, such as VSM, 5S, standardization, and cause-and-effect diagrams. Conversely, the service sector organizations tend to rely more on the Pareto diagram, standardization, cause-and-effect diagram, and flow charts.

The following discussion focuses on presenting and critically analysing some key trends in the field of GLSS. These emerging trends include Sustainable Lean-Six Sigma (SLSS), the role of digitalization in GLSS for achieving operational excellence and environmental performance, the nexus between the circular economy and GLSS, global supply chain operations, sustainability and GLSS, and the relationship between sustainable project management and the GLSS approach.

5.2.1 Incorporating Sustainability and Lean Six Sigma—Towards a Sustainable Lean-Six Sigma Approach

Sustainability has become critical for companies to achieve competitive advantage and social legitimacy, and SLSS—an approach that incorporates Lean Six Sigma (LSS) and sustainability practices (Swarnakar et al. 2023)—is increasingly recognized as a promising approach (Rana and Jani 2025) to simultaneously address social, economic, and environmental concerns in organizational operations. According to Dyllick and Hockerts (2022, p. 131), sustainability in the corporate context can be described as "meeting the needs of a firm's direct and indirect stakeholders (such as shareholders, employees, clients, pressure groups, communities, etc.), without compromising its ability to meet the needs of future stakeholders as well". Bansal (2005) argued that sustainability in a broader sense involves promoting three principles of sustainable development—environmental integrity, social equity, and economic prosperity. However, in the corporate context, environmental integrity can be ensured through corporate environmental management, social equity can be enhanced via embracing corporate social responsibility initiatives, and economic prosperity can be accomplished through value creation by offering high-quality sustainable goods and services to customers.

Due to the mounting pressure imposed by government and regulatory agencies, changing public opinion, and other social expectations regarding social, environmental, and ethical responsibilities, modern companies increasingly embed sustainability into their core processes and operational practices (Erdil et al. 2018; Rathi et al. 2022). Accordingly, reputational and litigation concerns propel companies to enhance environmental and social impacts throughout their inter- and intra-organizational operations and supply chain practices. To this end, SLSS can be instrumental in simultaneously promoting sustainability and operational performance (Parmar and Desai 2021); however, its role in fostering sustainable outcomes for companies has received little attention (de Freitas et al. 2017) in scholarly and practical domains. Thus, it can be argued that utilizing the SLSS approach without considering its linkages with sustainability could seriously undermine its potential in achieving sustainability-oriented innovations in the form of sustainable products or services and process improvement goals, better compliance with regulatory standards, and improved resource utilization of pristine natural resources.

Walter et al. (2023) explore research developments focusing on SLSS frameworks for organizational performance. While the authors argue that a SLSS approach contributes to organizational performance, many organizations lack competency and awareness of integrating sustainability with LSS in an operations management context. At the same time, the SLSS concept lacks theoretical maturity, suggesting more concerted efforts are required to develop the theoretical underpinnings of SLSS in a holistic fashion, incorporating the social dimension along with the economic, operational, and environmental pillars (e.g., Cherrafi et al. 2017; Siegel et al. 2019; Utama and Abirfatin 2023). Cherrafi et al. (2017) and Khana et al. (2020) further argued that developing an integrated SLSS framework is critical to balancing an organization's operational performance with improved social and environmental impacts.

Caiado et al. (2019) argued that LSS is a key solution to address sustainability issues, particularly when companies align their operational programmes with sustainability-oriented initiatives holistically, fostering improved environmental outcomes, operational competence, corporate reputation, and supplier relations. In this regard, companies should consider establishing sustainable performance management systems that correspond to the internal and external drivers influencing business performance and managing their environmental impacts. Teixeira et al. (2021) argue that while the lean manufacturing philosophy is increasingly recognized as a key approach to achieving operational gains and waste minimization, the current business environment demands companies to go beyond the traditional bottom line—profit maximization and economic outcomes—and consider social and environmental sustainability aspects owing to growing stakeholder pressure. Nonetheless, the authors noted that most studies focus on sustainability's economic and environmental aspects, and the social aspects, such as occupational health and safety, human rights and labour practices, and community outreach and impacts, are the least examined in the emerging literature (Cherrafi et al. 2016; Ghaseminbojd et al. 2025; Rathi et al. 2022). Further, some authors have noted that the notion of SLSS has received little scholarly attention in the emerging literature (e.g., Cherrafi et al. 2016). At the same time, companies still need to learn more about successfully incorporating sustainability and LSS approaches into their operational management activities using suitable implementation frameworks (Cherrafi et al. 2016; Erdil et al. 2018; Kaswan and Rathi 2020; Rathi et al. 2022).

Similarly, Ghaseminbojd et al. (2025) investigated the role of GLSS in enhancing sustainability impacts. The authors reported that most studies conducted with this aim are classified as being conceptual in nature, focusing on the manufacturing sector, or considered limited in terms of offering holistic methodologies or standardized sustainability metrics. In particular, the authors argued that one of the critical issues in GLSS research is the paucity of standardized metrics that simultaneously evaluate economic, social, and environmental pillars of sustainability. Thus, it is imperative to develop a balanced and more comprehensive approach to GLSS that considers the traditional aspects of performance measurement and the aspects related to broader sustainability objectives. Similar views are expressed by Kaswan and Rathi (2021), who asserted that the emerging body of knowledge has paid little attention to sustainability improvement measures with suitable GLSS implementation frameworks. Consequently, Gandhi and Thanki (2025) developed a sustainability index by integrating lean, green, and Six Sigma tools. The proposed framework involving various indicators provides insights into how pro-sustainability operational practices can enhance economic, social, and environmental performance outcomes. For example, the sustainability indicators include environmental performance indicators such as solid waste, energy usage, hazardous waste, and gaseous emissions as well as social sustainability indicators such as employee turnover index, number of workplace injuries, and number of trained workers. Meanwhile, the economic sustainability indicators include operational cost, environmental cost, customer returns, and defect rate. Swarnakar et al. (2023) studied critical success factors for SLSS in the healthcare sector. The factors identified in the study are instrumental for prioritizing and deploying SLSS in achieving improved healthcare service delivery. Similarly, Swarnakar et al. (2025) developed another framework for the step-by-step deployment of SLSS in manufacturing. The framework comprises different stages such as selecting the LSS project, defining the problem, measuring the data, mapping the current process, analysing the firm's sustainability level, improving through alternative solutions, and continuous monitoring to control and sustain the process. The authors argue that the framework is helpful in guiding practitioners seeking to implement the integrated SLSS approach to achieve improved sustainable outcomes in the manufacturing environment.

5.2.2 Green-Lean-Six Sigma, Technology, and Digitalization

Digitalization of organizational production and operations management is emerging as a new normal, and many organizations are embracing advanced technologies to achieve operational efficiencies, competitiveness, and sustainable business outcomes. Vaidya et al. (2018) argued that rapid adoption of advanced technologies drives productivity. Moreover, Kaswan et al. (2023a) argued that organizations must embrace advanced technologies to stay competitive in an ever-changing, complex, uncertain, and evolving business environment where the need for information exchange and stakeholder expectations around social and environmental criteria continue to grow. While GLSS-I4.0 is in early stages of development (Kaswan et al. 2023a), combining GLSS and Industry 4.0 approaches is one of the potential solutions to respond to these emerging trends and stakeholder requirements. According to Kagermann (2014, p. 23), "digitization—the continuing convergence of the real and the virtual worlds will be the main driver of innovation and change in all sectors of our economy". Technological advancements, including digitization, are vital in implementing the Green-Lean-Six Sigma (GLSS) approach and enhancing organizational operational and sustainability performance. The emergence of Industry 4.0, mainly fuelled by technological improvements in the industrial sector (Dallasega et al. 2018), has encouraged scholars to explore the potential of incorporating traditional production operations management practices with emergent technologies (Dalenogare et al. 2018; Madzík et al. 2025; Shahin et al. 2020) to foster more dynamic process optimization and efficient real-time decision making. Nonetheless, in-depth empirical studies exploring the role of advanced technologies in GLSS are scarce.

Industry 4.0, also known as the Fourth Industrial Revolution, involves the transformation of industrial processes, including production and distribution of goods and services, through the intelligent networking of machines and processes enabled by information and communication technology. The Organisation for Economic Co-operation and Development (OECD) (2017, p. 15) stated that the digital technologies within Industry 4.0 are increasingly causing disruptions to organizations and industrial sectors, and this shift "will have far-reaching consequences for productivity, employment, skills, income distribution, trade, well-being and the environment", indicating that competitive pressure on companies to adopt these technologies would continue to rise in the coming years.

The notion of Industry 4.0 was pioneered by a German collaborative initiative between universities and the private sector (Kagermann 2014). Hermann et al. (2016) identified three key elements of Industry 4.0: smart factories, cyber-physical systems (CPS), and the Internet of Things (IoT). Industry 4.0 aimed at enhancing decision making, efficiency, productivity, and responsiveness, streamlining advanced production systems (Becker et al. 2017; Kagermann 2014; Khanzode et al. 2023), developing new products (Dalenogare et al. 2018), and realizing sustainable business outcomes (e.g., Bai et al. 2020; de Sousa Jabbour et al. 2018b; Ghobakhloo 2020; Javaid et al. 2022). Industry 4.0 builds on the concept of advanced manufacturing, also referred to as smart manufacturing, which involves a flexible system that automatically regulates production processes for diverse types of products and changing conditions (Schuh et al. 2017; Wang et al. 2016).

According to Frank et al. (2019), Industry 4.0 involves a new industrial phase in which a range of technologies are converging, intending to deliver digital solutions. Industry 4.0 is broadly conceived as "a system that gathers and analyzes data from the floor to make intelligent decisions in an automated manner" (Ahuett-Garza and

Kurfess 2018, p. 60). Lu (2017, p. 3) defines Industry 4.0 as "an integrated, adapted, optimized, service-oriented and interoperable manufacturing process which correlates with algorithms, big data and high technologies". Industry 4.0 is an umbrella concept associated with the application of a range of technologies in production and supply chain operations management contexts, including big data analytics, robotics, 3-D printing, Internet of Things (IoT), blockchains, CPS, machine learning, additive manufacturing, Artificial Intelligence (AI), and digital value chains (e.g., Ahuett-Garza and Kurfess 2018; Frank et al. 2019; Haleem and Javaid, 2019; Koh et al. 2019; Strandhagen et al. 2017). Ghobakhloo (2018) revealed that Industry 4.0 involves 12 design principles and 14 emerging technology trends in the manufacturing context. The author argued that manufacturing companies need to embrace Industry 4.0 sooner rather than later, as the application of such technologies will become increasingly important in the coming years. While Industry 4.0 technologies are crucial in terms of how business activities and processes are planned and executed, the full potential of these technologies depends on the extent to which they are incorporated into operations and supply chain systems (Ahuett-Garza and Kurfess 2018).

Recent studies suggest that integrating GLSS with Industry 4.0 (e.g., Kaswan et al. 2023a; Skalli et al. 2023, 2025; Sordan et al. 2022) can be instrumental in advancing operational excellence and sustainability. However, with a few exceptions (e.g., Ibrahim and Kumar 2024; Mahajan et al. 2025), most existing research is either theoretical, conceptual, or literature reviews that provide initial insights into how Industry 4.0 technologies can be incorporated into GLSS and its related aspects (e.g., Citybabu and Yamini 2024; Letchumanan et al. 2022; Pongboonchai-Empl et al. 2024; Tissir et al. 2023).

Kaswan et al. (2023a) developed an integrative GLSS framework combining Industry 4.0 and GLSS concepts, aimed at enhancing organizational sustainability performance. The authors revealed that combining GLSS and Industry 4.0 involves common aspects in terms of enablers and inhibiting factors, and the integrated application of selected GLSS and Industry 4.0 tools and techniques helps mitigate environmental emissions, making industrial processes automated, streamlined, and responsive, which in turn fosters sustainability performance. In this regard, the authors combined Industry 4.0 practices with GLSS tools including SIPOC, project charters, and voice of the customer (VOC) with IoT tools such as barcodes, scanners, and radio frequency identification (RFID) to identify suitable GLSS projects. Similarly, other GLSS tools such as fishbone diagram, FMEA, and 5 whys are integrated with CPS and IoT to determine the potential root causes of waste and inefficiencies. In addition, DOE, EVSM, LCA, and Pugh matrix were combined with cloud manufacturing, additive manufacturing, VSM 4.0, and CPS to identify and implement the potential solutions. Following this, the authors emphasized integrating LCA and EVSM with IoT, CPS, and VSM 4.0 to control and sustain the adopted solutions. However, the authors acknowledged that an integrated GLSS-Industry 4.0 approach requires organizations to invest in training and development initiatives, and technology infrastructure should be developed to reap the potential synergies offered by simultaneously implementing these distant but related approaches. In another study,

Ooi et al. (2023) examined the effects of lean production on sustainable organizational performance with the moderating role of Industry 4.0 technologies adoption. Their findings demonstrated that Industry 4.0 adoption positively moderates the relationship between lean production and sustainable organizational performance.

Sahoo and Upadhyay (2024) explored the role of Industry 4.0, LSS, and circular supply chain management practices in promoting sustainability (i.e., triple bottom line) performance. The findings revealed that Industry 4.0 adoption helps create leaner and more sustainable processes. The authors argued that while Industry 4.0 technologies foster sustainability, successful implementation largely depends on vigilant planning, vigorous infrastructure, and organizational commitment. Accordingly, organizations need to address critical issues such as data privacy and security, skills inadequacy, and employee resistance to change to obtain the desired advantages of Industry 4.0 technologies, including value creation for stakeholders and building a sustainable future. Mahajan et al. (2025) explored the critical factors that impede Industry 4.0 adoption in promoting lean, agile, resilient, and green practices in the agri-food supply chain context. Their findings showed that lack of top management commitment and understanding, lack of long-term vision, and inadequate incentives and government support were the most significant barriers to adopting Industry 4.0 technologies.

Big data (e.g., Belhadi et al. 2023, 2020; Gupta et al. 2020), and other advanced technologies such as additive manufacturing, CPS, and the IoT (e.g., Tissir et al. 2023) contribute to organizational operational excellence. For instance, additive manufacturing allows efficient production and reduction of natural resource wastage, which could enhance positive impacts on the natural environment (Kaswan et al. 2023a). Similarly, technologies such as virtual and augmented reality, autonomous robots, and sensor-enabled technical systems are helpful in promoting social sustainability and an inclusive human-centred work environment by reducing employees' mental stress and physical strain, promoting employee training and development and collaboration, and enhancing wellbeing, safety, and job satisfaction (e.g., Kaswan et al. 2023a). Strandhagen et al. (2017) suggested that applying CPS, obtaining real-time insights from big data analytics, and embracing smart processes enable 'choice editing'—removing products unsafe for consumption (customer orientation and safety)—and achieve improved social and environmental performance, such as emission reduction by optimizing transportation routes, customer service, and health-related benefits. Nevertheless, there is still a gap in implementing such technologies in industrial systems and processes (Frank et al. 2019) and also linking them to wider sustainability outcomes.

According to Anthony et al. (2018, p. 1041), while linkages exist between LSS approaches and big data and hold substantial "potential to make production workflows more customer-centric than ever before", such linkages are not thoroughly investigated in emerging theoretical and empirical studies. Fayyaz et al. (2025) noted that big data analytics can be valuable in terms of providing real-time information that fosters a firm's decision-making capability and economic performance. The authors suggested that improved implementation of big data analytics enables companies to align their strategies with GLSS and sustainable supply chain management, which in turn leads to superior environmental performance. In another study, Khattab (2025) demonstrated that big data analytics adoption is positively associated withLSS, green production, and environmental performance.

Gupta et al. (2020) investigated the application of big data analytics techniques in various stages of LSS. Big data analytics deals with collecting, organizing, and analysing large data sets to unveil hidden patterns and reveal other useful information (Heymann and Rodier, 2003), which can be utilized for improved business decision making. The application of big data analytics in the LSS domain is an emerging phenomenon. Anthony et al. (2018) posited that companies need to align their internal practices and systems with the external environment to achieve competitiveness in the marketplace. To this end, the incorporation of big data analytics and LSS can be helpful in environmental scanning and for intelligence purposes, i.e., what is happening in the external environment of the firm, which in turn could help managers to devise suitable actions and strategies to respond to market changes with a more informed and factual approach.

Belhadi et al. (2023) developed a framework integrating big data analytics-driven GLSS to improve environmental outcomes. The authors suggested that big data analytics tools and techniques could be instrumental in providing real-time quality control, event-based inspection, predictive maintenance, and overall environmental performance. In a similar vein, Kaswan et al. (2023a) suggested that big data provides a broad spectrum of data concerning process parameters, and big data analytics helps in the selection of optimum parameters and provides real-time insights in terms of reducing waste and energy use, managing inefficiencies, and supporting informed decision-making necessary for achieving sustainable business outcomes.

5.2.3 Green-Lean-Six Sigma and the Circular Economy

The circular economy notion, generally conceived as a circular closed-loop system, is increasingly gaining traction in contemporary industrial operations and supply chain management scholarly and practitioner domains. While the circular economy is not new, its adoption in industrial processes started in the 1970s (Ellen MacArthur Foundation 2012). Early scholars such as Cooper (1999) also criticized the linear economic system, which assumes infinite natural resources for industrial use and a limitless ecosystem capacity to absorb waste and pollution generated from production and consumption activities. Nonetheless, the notion gained momentum and started garnering global recognition among current scholars and practitioners, which can be attributed to the Ellen MacArthur Foundation's report entitled "Towards the Circular Economy: Opportunities for the Consumer Goods Sector", which was instrumental in advancing its importance and highlighting the economic rationale for the corporate sector (e.g., Ellen MacArthur Foundation 2012).

The Ellen MacArthur Foundation's report identified the inherent limitations of the traditional, linear production model—take, make, and dispose—and the significance of transforming business processes and industrial systems to a more restorative and regenerative nature-friendly model, aligned with the sustainable development agenda. One such model, the 9R framework proposed by Potting et al. (2017), identified key circularity strategies within production chain phases that could help companies transition from a linear economy to a circular economy. The authors noted that a higher level of circularity enables utilization of less natural resources and minimal impacts on the natural environment. Recycling and recovery are considered low-circularity strategies, whereas repurposing, remanufacturing, refurbishing, repairing, and reusing strategies can extend the lifespan of products and their parts. More advanced circular strategies, such as reducing, rethinking, and refusing, foster innovative ways to manufacture and use products. Further, the systems perspective of the circular economy provides macro, meso, and micro-level applications of the circular economy. The macro-systems perspective of the circular economy considers the global, regional (e.g., Kristensen and Mosgaard 2020), or "industrial composition and structure of the entire economy" (Kirchherr et al. 2017, p. 224). The meso-systems perspective of the circular economy focuses on eco-industrial parks (e.g., Gómez et al. 2018; Zhang et al. 2010), while micro-systems look at firm, consumer, and product-level circularity initiatives (e.g., Kristensen and Mosgaard 2020).

Many scholars have argued that GLSS tools and practices are aligned with the circular economy approach (e.g., Farrukh et al. 2023; Shrivastava and Mishra 2025a, 2025b). Further, the circular economy concept operationalizes sustainable development for the business sector, suggesting how businesses can effectively contribute to society, enhancing their impact on the natural environment, communities, and other relevant stakeholders (Kirchherr et al. 2017; Murray et al. 2017). Murray et al. (2017) further noted that the importance of the circular economy lies in its potential adoption and engagement by all types of companies to address the sustainable development agenda.

One of the prominent definitions of circular economy is provided by the Ellen MacArthur Foundation (2012, p. 7), which states:

> Circular economy is an industrial system that is restorative or regenerative by intention and design. It replaces the 'end-of-life' concept with restoration, shifts towards the use of renewable energy, eliminates the use of toxic chemicals, which impair reuse, and aims for the elimination of waste through the superior design of materials, products, systems, and, within this, business models.

Alternatively, scholars such as Kirchherr et al. (2017, p. 224) conceptualize the circular economy notion as:

> An economic system that is based on business models which replace the 'end-of-life' concept with reducing, alternatively reusing, recycling and recovering materials in production/distribution and consumption processes, thus operating at the micro level (products, companies, consumers), meso level (eco-industrial parks) and macro level (city, region, nation and beyond), with the aim to accomplish sustainable development, which implies creating environmental quality, economic prosperity and social equity, to the benefit of current and future generations.

One of the concerns raised by scholars about the circular economy is the inadequate focus on the social dimensions of sustainable development (e.g., Murray et al. 2017). The notion of sustainability emphasizes the significance of incorporating the social, economic, and environmental aspects into business operations and strategic decisions (Ajmal et al. 2018). Similarly, the World Commission on Environment and Development (1987, p. 43) defines sustainable development as "development which meets the needs of the present without compromising the ability of future generations to meet their own needs". This definition recognizes the significance of intergenerational equity and concerns related to social justice, which are not explicitly studied in the circular economy literature and thus need further exploration. Building on this limited focus on the social dimension while conceptualizing circular economy, Murray et al. (2017, p. 369) put forward an updated definition of circular economy as "an economic model wherein planning, resourcing, procurement, production and reprocessing are designed and managed, as both process and output, to maximize ecosystem functioning and human well-being".

Ghaseminbojd et al. (2025) state that considerable potential exists for integrating GLSS with a circular economy concept from a sustainability perspective. Kalemkerian et al. (2022) argued that green lean tools and a circular economy strategy overlap in terms of addressing waste issues; however, the green lean strategy is limited when it comes to re-utilizing waste in the industrial process, which can be complemented through embracing circular economy principles. The authors further posited that "the green lean approach has been mainly developed as an extension of the traditional lean philosophy, which is based on a linear mode of production focused on waste-reduction activities. As a result, the value creation focus of lean thinking has been redirected towards the waste reduction aim of green lean thinking" (Kalemkerian et al. 2022, p. 1073). Thus, the traditional concept of 'value' from a GLSS perspective should be re-conceptualized towards embracing more inclusive circular and regenerative value creation principles that could lead to optimal sustainability outcomes.

A study by Farrukh et al. (2023) revealed that GLSS practices foster companies' transition to a circular economy not only through waste and emission reduction but also through waste management such as efficient collection and sorting of process waste, resource circulation by reusing and recycling waste for manufacturing new products, and environmental conservation by minimum use of raw materials, avoiding hazardous materials, and using sustainable and bio-based materials. The authors also determined the enabling factors such as strategic (e.g., leadership commitment and support), technical (e.g., innovation and technical competency), human resource (employee empowerment and development), and external stakeholders (e.g., government support and supply chain collaboration), contribute to GLSS adoption in promoting the circular economy concept. Similarly, a study by de Sousa Jabbour et al. (2018a) revealed that Industry 4.0 fosters achieving a circular economy. Thus, it can be argued that adopting advanced digital technologies could be instrumental in GLSS implementation, which helps enhance the circularity of resources with industrial production and supply chain operations.

Sahoo and Upadhyay (2024) investigated the advantages of incorporating Industry 4.0 technologies, LSS practices, and circular supply chain management to achieve a firm's triple bottom line (TBL) performance. The analysis of the findings revealed that Industry 4.0 directly and significantly affects economic performance only. The relationship between Industry 4.0 and environmental performance is determined by the implementation of LSS practices and a circular supply chain management approach. For social performance, the impact of LSS practices was found to be insignificant, whereas circular supply chain management mediates this relationship. Finally, in relation to the economic performance dimension, LSS practices and circular supply chain management partially mediate between the Industry 4.0 deployment and a firm's TBL performance. Thus, it can be argued that embracing Industry 4.0 technologies enhances LSS practices and circular supply chain management performance, which in turn helps improve the firm's TBL performance.

Skalli et al. (2024) explored the role of Circular Lean Six Sigma 4.0 (CLSS 4.0) in promoting manufacturing sustainability. The authors identified drivers and barriers that could help manufacturing companies adopt an integrated CLSS 4.0 approach. The analysis of the findings demonstrates that operational and environmental performance, increasing customer requirements, gaining a competitive advantage, and market growth are the main implementation drivers for integrated CLSS 4.0 in the manufacturing companies' context. Conversely, insufficient organizational resources (both tangible and intangible), data privacy, technical issues, and management support act as barriers to implementing the CLSS approach. Similarly, in another study, Skalli et al. (2024) investigated the relationship among LSS, the circular economy, and Industry 4.0 technologies, and their effect on sustainable organizational performance within Moroccan manufacturing firms. The results revealed that LSS, the circular economy, and Industry 4.0 technologies positively affect firms' sustainable organizational performance. Thus, drawing on these findings, it can be argued that practitioners need to pay careful attention to the selected approaches to achieve improved sustainable business performance.

5.2.4 Green-Lean-Six Sigma and Global Supply Chain Sustainability

Organizations increasingly entrench sustainability principles into modern supply chain operations to achieve social legitimacy and competitiveness. It is becoming increasingly important that organizations understand the broader impacts of business on society, and for that purpose, it is vital to consider not only the intra-organizational operations but also the environmental and social impacts associated with extended inter-organizational supply chain operations (Sajjad, 2015). Supply chain sustainability management is a broad concept that refers to "the management of material, information and capital flows as well as cooperation among companies along the supply chain while taking goals from all three dimensions of sustainable development, i.e., economic, environmental and social, into account which are derived from customer and stakeholder requirements" (Seuring and Müller, 2008, p. 1700).

Incorporating sustainability into supply chain operations is critical from multiple perspectives, including improved risk and stakeholder relationship management, development of innovative green products, sustainable service delivery, reputation enhancement, building loyalty and goodwill among customers, compliance with regulations, and overall superior financial, social, and environmental outcomes (e.g., Golicic and Smith, 2013; Wang and Dai 2018; Zailani et al. 2012). Given the significance and wide range of benefits of embedding sustainability into supply chain operations, it is critical to consider how GLSS practices are linked to and support fostering sustainability across all phases (upstream, midstream, and downstream) of extended global supply chain operations.

While limited research explores how the GLSS approach can be integrated with sustainable supply chain management (e.g., Araújo Barbosa et al. 2023; Sreedharan and Raju 2018), investigating the relationship between these approaches could be useful to advance the sustainability agenda from operations and supply chain perspectives. It is worth noting that most research studies examined limited aspects of GLSS, such as lean, with reference to sustainable supply chain approaches, such as green supply chain management. However, GLSS and sustainable supply chain management are rarely researched holistically, and most studies looked at the relationship between lean practices and a green supply chain management approach (e.g., Aruleswaran et al. 2025; Gál et al. 2025; Long et al. 2025; Silva and Gomes 2025; Singh et al. 2020; Sreedharan and Raju 2018; Zaidi and Lakhal 2025), indicating the social aspects were least considered in prior studies, both from GLSS and sustainable supply chain management perspectives. On the other hand, limited studies investigated GLSS application in different stages of sustainable supply chain management. In this regard, most of the existing studies examined GLSS application in mid-stream supply chain operations such as manufacturing and production operations (McGlinchey et al. 2025; Rahardjo et al. 2025; Sharma and Singh 2025) while neglecting the upstream and downstream activities including sustainable procurement, warehousing, logistics, and distribution aspects. Further, none of the existing studies examined GLSS application in reverse logistics from a sustainable and circular supply chain management perspective.

Fayyaz et al. (2025) studied the relationship between big data analytics, GLSS, and sustainable supply chain management. The findings revealed that combining big data analytics, GLSS, and sustainable supply chain management enhances firms' sustainability performance and competitiveness. Araújo Barbosa et al. (2023) conducted a systematic literature review of lean, Six Sigma, and sustainability case studies on supply chain management. The review's findings demonstrated that a few studies explored logistics and LSS linkages. This scarcity is even more pronounced regarding the relationship between GLSS and the context of supply chain operations. However, their findings identified a range of GLSS tools that can be utilized to enhance sustainable supply chain performance. Thus, it can be argued that integrating sustainable supply chain management and GLSS should be further examined in future research endeavours, as there are diverse potential unexplored avenues and opportunities. Unveiling these research pathways in an integrated way, capturing relevant sustainable supply chain management and GLSS approaches can lead to significant payoffs in promoting the global sustainable development agenda, not only at the intra-organizational levels (i.e., from an internal operations perspective) but beyond a firm's boundaries (external supply chain level—upstream and downstream), where untapped social, economic, and environmental performance enhancement opportunities exist.

5.2.5 *Green-Lean-Six Sigma and Sustainable Project Management*

The integration of GLSS is gaining traction in sustainable project management and delivery. Projects are considered a valuable means to promote sustainable business performance (Kivilä et al. 2017) by developing sustainable products and making operational processes more efficient, environmentally, and socially sustainable. Sustainable project management is broadly conceived as integrating social, environmental, and economic aspects of sustainability into project management practices (Shaukat et al. 2022). Silvius and Schipper (2014) defined a sustainable project management approach as:

> The planning, monitoring and controlling of project delivery and support processes, with consideration of the environmental, economic and social aspects of the life cycle of the project's resources, processes, deliverables and effects, aimed at realizing benefits for stakeholders, and performed in a transparent, fair and ethical way that includes proactive stakeholder participation (p. 79).

Silvius and Schipper's definition of sustainable project management considers salient sustainability dimensions—social, economic, and environmental aspects with reference to stakeholder expectations concerning transparency, ethics, and fair business practices—and core elements of project management, including superior project delivery and better support processes throughout a project life cycle. Sustainable project management includes two main perspectives—'sustainability of projects' and 'sustainability by projects'—each with a distinctive focus (e.g., Cillo et al. 2019; Khalifeh et al. 2020; Silvius et al. 2024; Zwikael and Huemann 2023). Sustainability of projects focuses on improving the 'process-related aspects' by which the project is executed (i.e., embracing sustainability practices in project management), such that its sustainability is ensured throughout the lifecycle stages and the project is managed in the most effective and efficient way without overexploiting pristine natural resources or producing damaging environmental or social impacts. Meanwhile, the 'sustainability by project' standpoint considers the project's enduring impacts and the project as an essential vehicle for achieving sustainable development to foster substantial long-term environmental, social, and economic benefits (Williams et al. 2020).

The emerging literature linking sustainable project management and GLSS appears to focus more on the 'sustainability of project' rather than the 'sustainability by project'. Thus, it makes sense for scholars to connect GLSS (a process-oriented approach) with sustainable practices in the context of project management. For instance, Kaswan et al. (2023b) examined the role of sustainability-focused GLSS project selection for the manufacturing industry in a complex decision-making environment to develop a GLSS implementation framework for improved organizational performance. This study focused on the conceptualization and planning stages of the GLSS project life cycle, indicating the demonstration of the 'sustainability of project' perspective. Perez-Burgoin et al. (2024) argued the necessity for committed and leadership trained in implementing GLSS projects in organizations, driving sustainability benefits. Antony et al. (2020) investigated the antecedents of GLSS project failure in another study. The findings revealed that a lack of management commitment, employee resistance to change, poor rewards and recognition systems, inconsistent monitoring and control of the projects, and inadequate communication are the leading causes of LSS project failure. Based on the above discussion, it can be argued that there is considerable scope for expanding the existing research focus to 'sustainability by project' along with 'sustainability of project'. This renewed scholarly focus could help researchers explore new research pathways and interesting intellectual avenues that could advance GLSS scholarship—empirically exploring the role of GLSS in making significant contributions towards developing sustainable outcomes in the form of sustainable products and service delivery with significant impacts on the natural environment and society.

5.3 Summary

This chapter examined the emerging trends transforming the application of GLSS in manufacturing and service organizations. It highlighted various developing trends influencing the GLSS phenomenon, including a shift from GLSS (economic and environmental) towards SLSS, a more holistic approach incorporating triple bottom line aspects.

Next, the chapter discussed the interplay between GLSS and advanced technologies (e.g., Industry 4.0 and digitization), which can be instrumental in fostering operational excellence and sustainability. After that, a critical analysis of the concept of a circular economy was discussed in relation to the GLSS strategy. Finally, the chapter concluded with an in-depth exploration of the role of GLSS in global supply chain operations and modern project management.

References

Ahmed, S., Hawarna, S., Alqasmi, I., Mohiuddin, M., Rahman, M. K., & Ashrafi, D. M. (2024). Role of Lean Six Sigma approach for enhancing the patient safety and quality improvement in the hospitals. *International Journal of Healthcare Management*, *17*(1), 52–62.

Ahuett-Garza, H., & Kurfess, T. (2018). A brief discussion on the trends of habilitating technologies for industry 4.0 and smart manufacturing. *Manufacturing Letters*, 15, 60e63. https://doi.org/10.1016/j.mfglet.2018.02.011

Ajmal, M. M., Khan, M., Hussain, M., & Helo, P. (2018). Conceptualizing and incorporating social sustainability in the business world. *International Journal of Sustainable Development & World Ecology*, 25(4), 327–339.

Antony, J., Lizarelli, F. L., & Fernandes, M. M. (2020). *A global study into the reasons for lean six sigma project failures: key findings and directions for further research.* IEEE Transactions on Engineering Management, 69(5), 2399-2414.

Antony, J., Gupta, S., Sunder M, V., & Gijo, E. V. (2018). Ten commandments of Lean Six Sigma: a practitioners' perspective. *International Journal of Productivity and Performance Management*, 67(6), 1033–1044.

Araújo Barbosa, F. T., Santana Peruchi, R., Naomi Morioka, S., & Rotella Junior, P. (2023). Lean, six sigma and sustainability case studies on supply chain management: a systematic literature review. *GeSec: Revista de Gestao e Secretariado*, 14(9).

Aruleswaran, A., Muraliraj, J., & Zailani, S. (2025). Lean six sigma and sustainable supply chain management: a case study in electric vehicle parts manufacturing. *International Journal of Lean Six Sigma*. https://doi.org/10.1108/IJLSS-04-2024-0069

Bai, C., Dallasega, P., Orzes, G., & Sarkis, J. (2020). Industry 4.0 technologies assessment: a sustainability perspective. *International Journal of Production Economics*, 229, 107776.

Bansal, P. (2005). Evolving sustainably: A longitudinal study of corporate sustainable development. *Strategic Management Journal*, 26(3), 197-218.

Becker, T., Blocher, A., Bücker, I., Drewel, M., Faath, A., Harland, T., ... & Zeller, V. (2017). Industrie 4.0 maturity index [eng.]: managing the digital transformation of companies. Herbert Utz Verlag.

Belhadi, A., Kamble, S. S., Zkik, K., Cherrafi, A., & Touriki, F. E. (2020). The integrated effect of Big Data Analytics, Lean Six Sigma and Green Manufacturing on the environmental performance of manufacturing companies: The case of North Africa. *Journal of Cleaner Production, 252*, 119903.

Belhadi, A., Kamble, S. S., Gunasekaran, A., Zkik, K., M, D. K., & Touriki, F. E. (2023). A Big Data Analytics-driven Lean Six Sigma framework for enhanced green performance: a case study of chemical company. *Production Planning & Control*, 34(9), 767–790.

Caiado, R. G. G., Quelhas, O. L. G., Nascimento, D. L. D. M., Anholon, R., & Leal Filho, W. (2019). Towards sustainability by aligning operational programmes and sustainable performance measures. *Production Planning & Control*, 30(5-6), 413–425.

Cherrafi, A., Elfezazi, S., Chiarini, A., Mokhlis, A., & Benhida, K. (2016). The integration of lean manufacturing, Six Sigma and sustainability: A literature review and future research directions for developing a specific model. *Journal of Cleaner Production*, 139, 828–846.

Cherrafi, A., Elfezazi, S., Govindan, K., Garza-Reyes, J. A., Benhida, K., & Mokhlis, A. (2017). A framework for the integration of Green and Lean Six Sigma for superior sustainability performance. *International Journal of Production Research*, *55*(15), 4481–4515.

Cillo, V., Petruzzelli, A. M., Ardito, L., & Del Giudice, M. (2019). Understanding sustainable innovation: A systematic literature review. *Corporate Social Responsibility and Environmental Management*, 26(5), 1012–1025.

Citybabu, G., & Yamini, S. (2024). Lean Six Sigma and Industry 4.0–a bibliometric analysis and conceptual framework development for future research agenda. *International Journal of Productivity and Performance Management*, 73(5), 1502–1534.

Cooper, T. (1999). Creating an economic infrastructure for sustainable product design. *Journal of Sustainable Product Design*, 8, 7–18.

Dalenogare, L. S., Benitez, G. B., Ayala, N. F., & Frank, A. G. (2018). The expected contribution of Industry 4.0 technologies for industrial performance. *International Journal of Production Economics*, 204, 383–394.

Dallasega, P., Rauch, E., & Linder, C. (2018). Industry 4.0 as an enabler of proximity for construction supply chains: A systematic literature review. *Computers in Industry*, 99, 205–225.

de Freitas, J. G., Costa, H. G., & Ferraz, F. T. (2017). Impacts of Lean Six Sigma over organizational sustainability: A survey study. *Journal of Cleaner Production*, 156, 262–275.

de Sousa Jabbour, A. B., Jabbour, C. J. C., Godinho Filho, M., & Roubaud, D. (2018a). Industry 4.0 and the circular economy: a proposed research agenda and original roadmap for sustainable operations. *Annals of Operations Research*, 270(1), 273–286.

de Sousa Jabbour, A. B. L., Jabbour, C. J. C., Foropon, C., & Godinho Filho, M. (2018b). When titans meet–Can industry 4.0 revolutionise the environmentally-sustainable manufacturing wave? The role of critical success factors. *Technological Forecasting and Social Change*, 132, 18–25.

Dyllick, T., & Hockerts, K. (2002). Beyond the business case for corporate sustainability. *Business Strategy and the Environment*, 11(2), 130-141.

Ellen MacArthur Foundation (2012). Towards the circular economy: opportunities for the consumer goods, https://content.ellenmacarthurfoundation.org/m/50c85a620a58955/original/Towards-the-circular-economy-Vol-2.pdf

Erdil, N. O., Aktas, C. B., & Arani, O. M. (2018). Embedding sustainability in lean six sigma efforts. *Journal of Cleaner Production*, 198, 520-529.

Farrukh, A., Mathrani, S., & Sajjad, A. (2023). Green-lean-six sigma practices and supporting factors for transitioning towards circular economy: A natural resource and intellectual capital-based view. *Resources Policy*, 84, 103789.

Fayyaz, A., Liu, C., Xu, Y., Khan, F., & Ahmed, S. (2025). Untangling the cumulative impact of big data analytics, green lean six sigma and sustainable supply chain management on the economic performance of manufacturing organisations. *Production Planning & Control*, 36(8), 1137–1154.

Frank, A. G., Dalenogare, L. S., & Ayala, N. F. (2019). Industry 4.0 technologies: Implementation patterns in manufacturing companies. *International Journal of Production Economics*, 210, 15–26.

Gál, T., Fenyves, V., Csipkés, M., Terjék, L., Kovács, S., & Buglyó-Nyakas, E. (2025). Effects of lean and green supply chain management practices on the performance of Hungarian manufacturing companies. *Discover Sustainability*, 6(1), 1005.

Gandhi, J. D., & Thanki, S. (2025). Sustainability index development by integrating lean green and Six Sigma tools: a case study of the Indian manufacturing industry. *International Journal of Productivity and Performance Management*, 74(3), 793–818.

Ghasemibojd, F., Franchetti, M. J., & George, B. (2025). Green lean six sigma for sustainable development: a systematic review of evolution, challenges, and future pathways. *Clean Technologies and Environmental Policy*, 1–24.

Ghobakhloo, M. (2018). The future of manufacturing industry: a strategic roadmap toward Industry 4.0. *Journal of Manufacturing Technology Management, 29*(6), 910–936.

Ghobakhloo, M. (2020). Industry 4.0, digitization, and opportunities for sustainability. *Journal of Cleaner Production*. https://doi.org/10.1016/j.jclepro.2019.119869

Golicic, S. L., & Smith, C. D. (2013). A meta-analysis of environmentally sustainable supply chain management practices and firm performance. *Journal of Supply Chain Management*, 49(2), 78–95.

Gómez, A. M. M., González, F. A., & Bárcena, M. M. (2018). Smart eco-industrial parks: A circular economy implementation based on industrial metabolism. *Resources, Conservation and Recycling*, 135, 58–69.

Gupta, S., Modgil, S., & Gunasekaran, A. (2020). Big data in lean six sigma: a review and further research directions. *International Journal of Production Research, 58*(3), 947-969.

Haleem, A., & Javaid, M. (2019). Additive manufacturing applications in industry 4.0: a review. *Journal of Industrial Integration and Management*, 4(04), 1930001.

Hermann, M., Pentek, T., Otto, B. (2016). *Design principles for Industrie 4.0 scenarios*. In: IEEE the 49th Hawaii International Conference on System Sciences (HICSS), Hawaii, pp 3928–3937.

Heymann, D. L., & Rodier, G. (2003). SARS: a global response to an international threat. *Brown Journal of World Affairs*, 10(2), 185–197.

Ibrahim, A., & Kumar, G. (2024). Selection of Industry 4.0 technologies for Lean Six Sigma integration using fuzzy DEMATEL approach. *International Journal of Lean Six Sigma*, 15(5), 1025–1042.

Javaid, M., Haleem, A., Singh, R. P., Suman, R., & Gonzalez, E. S. (2022). Understanding the adoption of Industry 4.0 technologies in improving environmental sustainability. *Sustainable Operations and Computers*, 3, 203–217.

Jin, H., Qu, P., Zhou, Y., Qu, Q., & Li, H. (2025). Green Lean Six Sigma practice in the industrial and service sectors: From the systematic literature review to an integration diagram. *Production Planning & Control*, 36(3), 394-419.

Kagermann, H. (2014). Change through digitization—value creation in the age of Industry 4.0. In Management of permanent change (pp. 23–45). Wiesbaden: Springer Fachmedien Wiesbaden.

Kalemkerian, F., Santos, J., Tanco, M., Garza-Reyes, J. A., & Viles, E. (2022). Analysing the alignment between the green lean and circular strategies: towards a circular lean approach. *Journal of Manufacturing Technology Management*, 33(6), 1059-1079.

Kaswan, M. S., Rathi, R., Cross, J., Garza-Reyes, J. A., Antony, J., & Yadav, V. (2023a). Integrating green lean six sigma and industry 4.0: a conceptual framework. *Journal of Manufacturing Technology Management*, 34(1), 87–121.

Kaswan, M. S., Rathi, R., Garza-Reyes, J. A., & Antony, J. (2023b). Green lean six sigma sustainability–oriented project selection and implementation framework for manufacturing industry. International Journal of Lean Six Sigma, 14(1), 33–71.

Kaswan, M. S., & Rathi, R. (2020). Green Lean Six Sigma for sustainable development: Integration and framework. *Environmental Impact Assessment Review*, 83, 106396.

Kaswan, M. S., & Rathi, R. (2021). An inclusive review of Green Lean Six Sigma for sustainable development: readiness measures and challenges. *International Journal of Advanced Operations Management*, 13(2), 129–166.

Khalifeh, A., Farrell, P., Alrousan, M., Alwardat, S., & Faisal, M. (2020). Incorporating sustainability into software projects: a conceptual framework. *International Journal of Managing Projects in Business*, 13(6), 1339-1361.

Khana, M. P., Talibb, N. A., & Kowangc, T. O. (2020). The development of a sustainability framework via lean green six sigma practices in SMEs based upon RBV theory. *Development*, 12(5), 135-156.

Khanzode, A. G., Sarma, P. R. S., & Goswami, M. (2023). Modelling interactions of select enablers of Lean Six-Sigma considering sustainability implications: an integrated circular economy and Industry 4.0 perspective. *Production Planning & Control*, 34(10), 1020–1036.

Khattab, S. A. (2025). The Impact of Big-data Analytics on Environmental Performance: The Mediating Roles of Lean Six Sigma and Green Production. *Jordan Journal of Business Administration*, 21(4), 539–562.

Kirchherr, J., Reike, D., & Hekkert, M. (2017). Conceptualizing the circular economy: An analysis of 114 definitions. *Resources, Conservation and Recycling*, 127, 221–232.

Kivilä, J., Martinsuo, M., & Vuorinen, L. (2017). Sustainable project management through project control in infrastructure projects. *International Journal of Project Management*, 35(6), 1167-1183.

Koh, L., Orzes, G., & Jia, F. (2019). The fourth industrial revolution (Industry 4.0): technologies disruption on operations and supply chain management. *International Journal of Operations & Production Management*, 39(6/7/8), 817–828.

Kristensen, H. S., & Mosgaard, M. A. (2020). A review of micro level indicators for a circular economy–moving away from the three dimensions of sustainability?. *Journal of Cleaner Production*, 243, 118531.

Letchumanan, L. T., Gholami, H., Yusof, N. M., Ngadiman, N. H. A. B., Salameh, A. A., Štreimikienė, D., & Cavallaro, F. (2022). Analyzing the factors enabling green lean six sigma implementation in the industry 4.0 era. *Sustainability*, 14(6), 3450.

Long, Y., Yang, Y., Feng, T., Cheng, Y., & Farooq, S. (2025). Green supply chain integration, lean management, and manufacturers' performance: role of supply chain risk management. *Total Quality Management & Business Excellence*, 36(5-6), 369-394.

Lu, Y. (2017). Industry 4.0: A survey on technologies, applications and open research issues. *Journal of Industrial Information Integration*, 6, 1–10.

Madzík, P., Falát, L., Jayaraman, R., Sony, M., & Antony, J. (2025). Exploring research trends in Lean, Six Sigma and Lean Six Sigma methodologies through a hybrid artificial intelligence approach. *Production Planning & Control*, 1–25.

Mahajan, P. S., Raut, R., Cheikhrouhou, N., Yadav, V. S., & Ghoshal, S. (2025). Industry 4.0 adoption challenges in lean-agile-resilient-green agri-food supply chain. *International Journal of Lean Six Sigma*, 16(4), 890–917.

McGlinchey, Y., Iqbal, J., Trubetskaya, A., Thenarasu, M., & McDermott, O. (2025). Design for green lean six sigma to improve sustainability in the pharmaceutical industry—a case study. Production & Manufacturing Research, 13(1). https://doi.org/10.1080/21693277.2025.2461010

Murray, A., Skene, K., & Haynes, K. (2017). The circular economy: an interdisciplinary exploration of the concept and Application in a global context. *Journal of Business Ethics*, 140(3), 369–380.

OECD (2017). The next production revolution: implications for governments and business. OECD Publishing, Paris. https://www.oecd.org/en/publications/the-next-production-revolution_9789264271036-en.html

Ooi, L. L., Teh, S. Y., & Cheang, P. Y. S. (2023). The impact of lean production on sustainable organizational performance: the moderating effect of industry 4.0 technologies adoption. *Management Research Review*, 46(12), 1802–1836.

Parmar, P. S., & Desai, T. N. (2021). Ranking the solutions of Sustainable Lean Six Sigma implementation in Indian manufacturing organization to overcome its barriers. *International Journal of Sustainable Engineering*, 14(3), 304-317.

Perez-Burgoin, M., Baez-Lopez, Y., Limon-Romero, J., Tlapa, D., & García-Alcaraz, J. L. (2024). Enablers for green lean six sigma adoption in the manufacturing industry. *Journal of Manufacturing Technology Management*, 35(6), 1199-1225.

Pongboonchai-Empl, T., Antony, J., Garza-Reyes, J. A., Komkowski, T., & Tortorella, G. L. (2024). Integration of industry 4.0 technologies into Lean Six Sigma DMAIC: a systematic review. *Production Planning & Control*, 35(12), 1403–1428.

Potting, J., et al., 2017. Circular economy: measuring innovation in the product chain. https://www.pbl.nl/uploads/default/downloads/pbl-2016-circular-economy-measuring-innovation-in-product-chains-2544.pdf

Rahardjo, B., Winnyarto, S. T., & Tjendra, V. (2025). Integration of green lean six sigma and TRIZ methodology for sustainable green manufacturing: a case study in plywood industry. International Journal of Lean Six Sigma. https://doi.org/10.1108/ijlss-09-2024-0201

Rana, J. A., & Jani, S. Y. (2025). A structural framework to achieve operational excellence by adopting sustainable Lean Six Sigma and industry 4.0 technologies. *International Journal of Mathematical, Engineering and Management Sciences*, 10(4), 1080.

Rathi, R., Kaswan, M. S., Garza-Reyes, J. A., Antony, J., & Cross, J. (2022). Green Lean Six Sigma for improving manufacturing sustainability: Framework development and validation. *Journal of Cleaner Production*, 345, 131130.

Rathi, R., Kaswan, M. S., Antony, J., Cross, J., Garza-Reyes, J. A., & Furterer, S. L. (2023). Success factors for the adoption of green lean six sigma in healthcare facility: an ISM-MICMAC study. *International Journal of Lean Six Sigma*, *14*(4), 864–897.

Sahoo, S., & Upadhyay, A. (2024). Improving triple bottom line (TBL) performance: analyzing impacts of industry 4.0, lean six sigma and circular supply chain management. *Annals of Operations Research*, 1–32.

Sajjad, A. (2015). Embedding sustainability into supply chain management: a New Zealand perspective: a thesis presented in partial fulfilment of the requirement for the degree of Doctor of Philosophy in Management at Massey University, Albany, Auckland, New Zealand (Doctoral dissertation, Massey University).

Schuh, G., Anderl, R., Gausemeier, J., ten Hompel, M., & Wahlster, W. (2017). Industrie 4.0 maturity index. Managing the digital transformation of companies, 61.

Seuring, S., & Müller, M. (2008). From a literature review to a conceptual framework for sustainable supply chain management. *Journal of Cleaner Production*, 16(15), 1699-1710.

Shahin, M., Chen, F. F., Bouzary, H., & Zarreh, A. (2020). Frameworks proposed to address the threat of cyber-physical attacks to lean 4.0 systems. *Procedia Manufacturing*, 51, 1184–1191.

Sharma, D., & Singh, D. (2025). Green Lean six sigma barriers and efficacy in ceramic manufacturing industries: a PLS-SEM study.

Shaukat, M. B., Latif, K. F., Sajjad, A., & Eweje, G. (2022). Revisiting the relationship between sustainable project management and project success: The moderating role of stakeholder engagement and team building. *Sustainable Development*, 30(1), 58–75.

Shrivastava, A., & Mishra, R. P. (2025a). Deploying lean six sigma and industry 4.0 framework in an auto motive manufacturing organization for establishing circular economy. OPSEARCH. https://doi.org/10.1007/s12597-025-00952-8

Shrivastava, A., & Mishra, R. P. (2025b). Integrating Lean Six Sigma and Industry 4.0 for advancing the circular economy: a holistic research study. Quality management journal, 32(1), 2–14. https://doi.org/10.1080/10686967.2024.2429805

Siegel, R., Antony, J., Garza-Reyes, J. A., Cherrafi, A., & Lameijer, B. (2019). Integrated green lean approach and sustainability for SMEs: From literature review to a conceptual framework. *Journal of Cleaner Production, 240*, 118205.

Silva, G. M., & Gomes, P. J. (2025). Lean production, green supply chain management and environmental performance: a configurational perspective based on the Portuguese context. *International Journal of Lean Six Sigma*, 16(2), 518–541.

Silvius, G., & Huemann, M. (Eds.). (2024). *Research handbook on sustainable project management*. Edward Elgar Publishing.

Silvius, A. G., & Schipper, R. P. (2014). Sustainability in project management competencies: analyzing the competence gap of project managers. *Journal of Human Resource and Sustainability Studies*, 2(2), 40–58.

Singh, J., Singh, H., & Kumar, A. (2020). Impact of lean practices on organizational sustainability through green supply chain management–an empirical investigation. *International Journal of Lean Six Sigma*, 11(6), 1035-1068.

Skalli, D., Charkaoui, A., Cherrafi, A., Garza-Reyes, J. A., Antony, J., & Shokri, A. (2023). Industry 4.0 and Lean Six Sigma integration in manufacturing: a literature review, an integrated framework and proposed research perspectives. *Quality Management Journal*, 30(1), 16–40.

Skalli, D., Charkaoui, A., Cherrafi, A., Shokri, A., Garza-Reyes, J. A., & Antony, J. (2024). Analysis of factors influencing Circular-Lean-Six Sigma 4.0 implementation considering sustainability implications: an exploratory study. *International Journal of Production Research*, 62(11), 3890–3917.

Skalli, D., Cherrafi, A., Charkaoui, A., Chiarini, A., Shokri, A., Antony, J., ... & Foster, M. (2025). Integrating Lean Six Sigma and Industry 4.0: developing a design science research-based LSS4. 0 framework for operational excellence. *Production Planning & Control*, 36(8), 1060–1086.

Sordan, J. E., Oprime, P. C., Pimenta, M. L., Silva, S. L. D., & González, M. O. A. (2022). Contact points between Lean Six Sigma and Industry 4.0: a systematic review and conceptual framework. *International Journal of Quality & Reliability Management*, 39(9), 2155–2183.

Sreedharan V, R., G, S., & Raju, R. (2018). Development of a Green Lean Six Sigma model for public sectors. *International Journal of Lean Six Sigma*, 9(2), 238–255.
Strandhagen, J. O., Vallandingham, L. R., Fragapane, G., Strandhagen, J. W., Stangeland, A. B. H., & Sharma, N. (2017). Logistics 4.0 and emerging sustainable business models. *Advances in Manufacturing*, 5(4), 359–369.
Swarnakar, V., Bagherian, A., & Singh, A. R. (2023). Prioritization of critical success factors for sustainable Lean Six Sigma implementation in Indian healthcare organizations using best-worst-method. *The TQM Journal*, 35(3), 630–653.
Swarnakar, V., Singh, A. R., Antony, J., Kr Tiwari, A., & Cudney, E. A. (2025). A theoretical framework to deploy sustainable Lean Six Sigma and its empirical application in manufacturing. *Total Quality Management & Business Excellence*, 36(5-6), 527-552.
Teixeira, P., Sá, J. C., Silva, F. J. G., Ferreira, L. P., Santos, G., & Fontoura, P. (2021). Connecting lean and green with sustainability towards a conceptual model. *Journal of Cleaner Production*, 322, 129047.
Tissir, S., Cherrafi, A., Chiarini, A., Elfezazi, S., & Bag, S. (2023). Lean six sigma and industry 4.0 combination: scoping review and perspectives. *Total Quality Management & Business Excellence*, 34(3–4), 261–290.
Utama, D. M., & Abirfatin, M. (2023). Sustainable Lean Six-sigma: A new framework for improve sustainable manufacturing performance. *Cleaner Engineering and Technology*, 17, 100700.
Vaidya, S., Ambad, P., & Bhosle, S. (2018). Industry 4.0–a glimpse. *Procedia Manufacturing*, 20, 233–238.
Vashishth, A., Lameijer, B. A., Chakraborty, A., Antony, J., & Moormann, J. (2024). Implementing Lean Six Sigma in financial services: the effect of motivations, selected methods and challenges on LSS program-and organizational performance. *International Journal of Quality & Reliability Management*, 41(2), 509–531.
Walter, O. M. F. C., Paladini, E. P., Henning, E., & Kalbusch, A. (2023). Recent developments in sustainable lean six sigma frameworks: literature review and directions. *Production Planning & Control*, *34*(9), 830–848.
Wang, J., & Dai, J. (2018). Sustainable supply chain management practices and performance. *Industrial Management & Data Systems*, 118(1), 2–21.
Wang, S., Wan, J., Zhang, D., Li, D., & Zhang, C. (2016). Towards smart factory for industry 4.0: a self-organized multi-agent system with big data based feedback and coordination. *Computer Networks*, 101, 158–168.
WCED. (1987). Our common future. Oxford: Oxford University Press.
Williams, T., Vo, H., Bourne, M., Bourne, P., Cooke-Davies, T., Kirkham, R., ... & Valette, J. (2020). A cross-national comparison of public project benefits management practices–the effectiveness of benefits management frameworks in application. *Production Planning & Control*, 31(8), 644–659.
Yadav, V., Kumar, V., Gahlot, P., Mittal, A., Kaswan, M. S., Garza-Reyes, J. A., ... & Owad, A. A. (2024). Exploration and mitigation of green lean six sigma barriers: a higher education institutions perspective. *The TQM Journal*, *36*(7), 2132–2153.
Zaidi, A., & Lakhal, L. (2025). The impact of lean manufacturing practices on green supply chain management and corporate sustainable performance: evidence from manufacturing firms operating in Tunisia. *International Journal of Productivity and Performance Management*.https://doi.org/10.1108/IJPPM-07-2024-0448
Zailani, S., Jeyaraman, K., Vengadasan, G., & Premkumar, R. (2012). Sustainable supply chain management (SSCM) in Malaysia: A survey. *International Journal of Production Economics*, 140(1), 330-340.
Zhang, L., Yuan, Z., Bi, J., Zhang, B., & Liu, B. (2010). Eco-industrial parks: national pilot practices in China. *Journal of Cleaner Production*, 18(5), 504–509.
Zwikael, O., & Huemann, M. (2023). Project benefits management: Making an impact on organizations and society through projects and programs. *International Journal of Project Management*, 41(8), 102538.

Chapter 6
Concluding Remarks

6.1 Introduction

This book provides a holistic view of a Green-Lean-Six Sigma (GLSS) approach, combining three strategies: green manufacturing, lean manufacturing, and Six Sigma. The book presents GLSS drivers, enablers, tools, and barriers that help firms achieve operational and environmental performance. The key drivers include cost reduction, waste reduction, company image, regulatory requirements, and customer pressure (Cherrafi et al. 2017; Farrukh et al. 2021a, 2021b; Hariyani and Mishra 2022; Shokri et al. 2021). While several GLSS tools can help organizations achieve sustainable performance, the significant ones include VSM, SIPOC, EMS, LCA, 3R, cause-and-effect analysis, Pareto diagram, and SPC (Farrukh et al. 2022; McDermott et al. 2024; McGlinchey et al. 2025; Sony and Naik 2019; Yadav et al. 2021). Nevertheless, the successful deployment of GLSS depends on the essential factors enabling GLSS execution, such as top management commitment, resource allocation, operational and technical support, employee training and education, and stakeholder cooperation (Farrukh 2025; Farrukh et al. 2023b; Letchumanan et al. 2022; Perez-Burgoin et al. 2024; Rathi et al. 2023; Yadav et al. 2023).

Along with the above, this book discusses the theoretical frameworks of the GLSS approach, comprising various organizational theories such as the natural resource-based view, the institutional theory-based view, stakeholder theory, and the intellectual capital-based view (Farrukh et al. 2021b, 2023a). The discussion covers the theoretical constructs of GLSS (drivers, enablers, critical success factors, and barriers). In this regard, the theoretical underpinnings of the GLSS strategy through the above theories can guide scholars and practitioners in gaining an in-depth understanding of this strategy for achieving sustainable performance in organizations.

In addition, this book provides a global perspective and analysis of the state-of-the-art research on the GLSS approach and its application in contemporary organizations. It offers a broad view of the GLSS applications in developed and developing

A. Farrukh and A. Sajjad, *A Manager's Guide to Green-Lean-Six Sigma Application*,
https://doi.org/10.1007/978-3-032-21111-8_6

economies. The book includes case studies of applying GLSS strategy in manufacturing organizations, including both discrete and process industries, followed by cases from the service sector, such as healthcare and education, which can guide practitioners working in both manufacturing and service industries regarding applying GLSS (Gholami et al. 2020; Liu and Miao 2006; Vinnari 2024; Marrucci et al. 2020; Parmar and Desai 2020; Tiwari et al. 2020). Further, the book discusses real-world examples and cases of GLSS deployment in small and medium enterprises (SMEs) to achieve sustainable performance (Bai et al. 2018; Singh et al. 2021).

The book further examines recent advances and future perspectives in the field of GLSS, including the integration of digitalization (such as Industry 4.0 technologies), the circular economy, and sustainable project management (Kaswan et al. 2022; Letchumanan et al. 2022; Shrivastava and Mishra 2025a, 2025b; Skalli et al. 2024), which can be instrumental in guiding students, scholars, and researchers in industrial engineering, environmental management, operations management, and business sustainability.

6.2 Creating Sustainable Organizations Through a Holistic Green-Lean-Six Sigma Approach

Using a holistic approach to a GLSS strategy, organizations can achieve the United Nations (UN) Sustainable Development Goals (SDGs). For example, a GLSS strategy can enable firms to achieve SDG 12—responsible production and consumption—using a mix of GLSS tools focusing on waste reduction, resource optimization, natural resource conservation, emission reduction, and environmental safety (Farrukh et al. 2023b).

This book provides practical guidance to managers and practitioners working in various domains, including senior executives, functional managers, sustainability/environmental advisors, and training consultants. As such, the GLSS approach can enable consultants and corporate managers to achieve operational and environmental performance simultaneously. In this regard, lean Six Sigma and environmental consultants can enhance their operational and technical skills through understanding the benefits of the combined tools and practices of the GLSS approach. Further, consultancy organizations can develop GLSS teams, including experts and consultants with expertise on Lean Six Sigma, digital technologies, circular economy, and sustainability, to initiate GLSS projects addressing the organizations' sustainable development goals. In this regard, GLSS consultants can initiate GLSS projects with a mix of GLSS tools, including VSM, LCA, 5S, Poka-yoke, Kaizen, 3R, EMS, SIPOC, and SPC, integrating with advanced technologies (e.g., Industry 4.0 and digitization), which can be instrumental in fostering operational excellence and sustainability. Further, the concept of a circular economy provides a way forward to expanding the scope of the GLSS strategy beyond achieving operational excellence in internal organizational operations and extended global supply chain networks.

Regarding managerial implications, the GLSS strategy can assist corporate managers in understanding the benefits of an integrated GLSS approach for achieving environmental sustainability without compromising operational performance. Managers can utilize the GLSS strategy to address internal and external stakeholders' requirements regarding emission reduction, environmental safety, natural resource conservation, workplace safety, customer satisfaction, lead time reduction, product quality, waste reduction, and recycling. Moreover, an integrated approach can help managers comply with national and international environmental regulatory requirements and avoid environmental fines and penalties. While GLSS tools can enable practitioners to achieve sustainable performance, their successful execution relies on critical factors, including top management commitment, resource allocation, employee motivation and empowerment, and stakeholder collaboration. Accordingly, the critical success factors of GLSS, as highlighted in Chap. 2, can guide corporate managers in understanding the significance of the essential factors for successful deployment. Furthermore, the GLSS strategy can assist senior executives and corporate managers in integrating different management systems and certifications such as ISO 9001, ISO 50001, and ISO 45001. In this regard, a mix of GLSS tools can help practitioners by addressing the requirements of the above certifications, including quality, environmental safety, product safety, energy conservation, and workplace safety.

6.3 Concluding Comments

The GLSS strategy incorporates three distinct—green, lean, and Six Sigma—albeit interrelated, synergetic approaches that, when applied in combination, enable organizations to achieve waste minimization, effective utilization of resources, and improvements in process and operational activities. Thus, integrating green, lean, and Six Sigma can be viewed as a new prospect for manufacturing and service organizations to simultaneously enhance sustainability and business performance. Drawing on pertinent organizational theories, this book takes a holistic stance on the GLSS approach, offering novel insights into the need for GLSS in the corporate business sector, applying GLSS tools, techniques, barriers, and success factors that promote organizational sustainability.

This timely book provides in-depth insights into understanding the holistic nature of the GLSS approach and its linkages to business sustainability. With topical and current issues discussed, this book is a "must-have" for business practitioners, consultants, academics, students, and those interested in environmental sustainability management and business excellence in developing and developed economies.

References

Bai, C., Satir, A., & Sarkis, J. (2018). Investing in lean manufacturing practices: an environmental and operational perspective. *International Journal of Production Research*, 1–15. https://doi.org/10.1080/00207543.2018.1498986

Cherrafi, A., Elfezazi, S., Govindan, K., Garza-Reyes, J. A., Benhida, K., & Mokhlis, A. (2017). A framework for the integration of Green and Lean Six Sigma for superior sustainability performance. *International Journal of Production Research*, *55*(15), 4481-4515. https://doi.org/10.1080/00207543.2016.1266406

Farrukh, A., Mathrani, S., & Sajjad, A. (2022). Managerial perspectives on green-lean-six sigma adoption in the flexible packaging industry: empirical evidence from an emerging economy. *Journal of Manufacturing Technology Management*, *33*(7), 1232-1255.

Farrukh, A., Mathrani, S., & Sajjad, A. (2021a). A comparative analysis of green-lean-six sigma enablers and environmental outcomes: a natural resource-based view. *International Journal of Lean Six Sigma*. https://doi.org/10.1108/ijlss-05-2021-0095

Farrukh, A., Mathrani, S., & Sajjad, A. (2021b). A natural resource and institutional theory-based view of green-lean-six sigma drivers for environmental management. *Business Strategy and the Environment*, *31*(3), 1074-1090. https://doi.org/10.1002/bse.2936

Farrukh, A., Mathrani, S., & Sajjad, A. (2023a). An exploratory study of green-lean-six sigma motivators for environmental sustainability: Managerial insights from a developed and developing economy. *Business Strategy and the Environment*. https://doi.org/10.1002/bse.3412

Farrukh, A., Mathrani, S., & Sajjad, A. (2023b). Green-lean-six sigma practices and supporting factors for transitioning towards circular economy: A natural resource and intellectual capital-based view. *Resources Policy*, *84*. https://doi.org/10.1016/j.resourpol.2023.103789

Farrukh, A. (2025). A comparative exploratory study of the critical success factors of a green-lean-six sigma strategy. *Business process Management journal*, 1-35. https://doi.org/10.1108/BPMJ-09-2024-0862

Gholami, H., Jamil, N., Mat Saman, M. Z., Streimikiene, D., Sharif, S., & Zakuan, N. (2020). The application of Green Lean Six Sigma. *Business Strategy and the Environment*.

Hariyani, D., & Mishra, S. (2022). Drivers for the adoption of integrated sustainable green lean six sigma agile manufacturing system (ISGLSAMS) and research directions. *Cleaner engineering and technology*, *7*, 100449.

Kaswan, M. S., Rathi, R., Cross, J., Garza-Reyes, J. A., Antony, J., & Yadav, V. (2022). Integrating Green Lean Six Sigma and industry 4.0: a conceptual framework. *Journal of Manufacturing Technology Management*, *34*(1), 87–121. https://doi.org/10.1108/jmtm-03-2022-0115

Letchumanan, L. T., Gholami, H., Yusof, N. M., Ngadiman, N. H. A. B., Salameh, A. A., Štreimikienė, D., & Cavallaro, F. (2022). Analyzing the Factors Enabling Green Lean Six Sigma Implementation in the Industry 4.0 Era. *Sustainability*, *14*(6). https://doi.org/10.3390/su14063450

Liu, F., & Miao, Z. (2006). The application of RFID technology in production control in the discrete manufacturing industry. 2006 IEEE International Conference on Video and Signal Based Surveillance,

Marrucci, L., Marchi, M., & Daddi, T. (2020). Improving the carbon footprint of food and packaging waste management in a supermarket of the Italian retail sector. *Waste Manag*, *105*, 594-603. https://doi.org/10.1016/j.wasman.2020.03.002

McDermott, O., Moloney, C., Noonan, J., & Rosa, A. (2024). Green Lean Six Sigma in the food industry: a systematic literature review. *British Food Journal*, *126*(13), 455-469. https://doi.org/10.1108/bfj-01-2024-0100

McGlinchey, Y., Iqbal, J., Trubetskaya, A., Thenarasu, M., & McDermott, O. (2025). Design for green lean six sigma to improve sustainability in the pharmaceutical industry—a case study. *Production & Manufacturing Research*, *13*(1). https://doi.org/10.1080/21693277.2025.2461010

Parmar, P. S., & Desai, T. N. (2020). Evaluating Sustainable Lean Six Sigma enablers using fuzzy DEMATEL: A case of an Indian manufacturing organization. *Journal of Cleaner Production, 265*. https://doi.org/10.1016/j.jclepro.2020.121802

Perez-Burgoin, M., Baez-Lopez, Y., Limon-Romero, J., Tlapa, D., & García-Alcaraz, J. L. (2024). Enablers for green lean six sigma adoption in the manufacturing industry. *Journal of Manufacturing Technology Management, 35*(6), 1199-1225. https://doi.org/10.1108/JMTM-09-2023-0396

Rathi, R., Kaswan, M. S., Antony, J., Cross, J., Garza-Reyes, J. A., & Furterer, S. L. (2023). Success factors for the adoption of green lean six sigma in healthcare facility: an ISM-MICMAC study. *International Journal of Lean Six Sigma, 14*(4), 864-897. https://doi.org/10.1108/IJLSS-02-2022-0042

Shokri, A., Antony, J., Garza-Reyes, J. A., & Upton, M. (2021). Scoping review of the readiness for sustainable implementation of Lean Six Sigma projects in the manufacturing sector. *International Journal of Quality & Reliability Management, ahead-of-print*(ahead-of-print). https://doi.org/10.1108/ijqrm-08-2020-0261

Shrivastava, A., & Mishra, R. P. (2025a). Deploying lean six sigma and industry 4.0 framework in an auto motive manufacturing organization for establishing circular economy. *OPSEARCH*. https://doi.org/10.1007/s12597-025-00952-8

Shrivastava, A., & Mishra, R. P. (2025b). Integrating Lean Six Sigma and Industry 4.0 for advancing the circular economy: A holistic research study. *Quality management journal, 32*(1), 2–14. https://doi.org/10.1080/10686967.2024.2429805

Singh, M., Rathi, R., & Garza-Reyes, J. A. (2021). Analysis and prioritization of Lean Six Sigma enablers with environmental facets using best worst method: A case of Indian MSMEs. *Journal of Cleaner Production, 279*. https://doi.org/10.1016/j.jclepro.2020.123592

Skalli, D., Charkaoui, A., Cherrafi, A., Garza-Reyes, J. A., Antony, J., & Shokri, A. (2024). Analyzing the integrated effect of circular economy, Lean Six Sigma, and Industry 4.0 on sustainable manufacturing performance from a practice-based view perspective. *Business Strategy and the Environment, 33*(2), 1208–1226.

Sony, M., & Naik, S. (2019). Green Lean Six Sigma implementation framework: a case of reducing graphite and dust pollution. *International Journal of Sustainable Engineering*, 1–10. https://doi.org/10.1080/19397038.2019.1695015

Tiwari, P., Sadeghi, J. K., & Eseonu, C. (2020). A sustainable lean production framework with a case implementation: Practice-based view theory. *Journal of Cleaner Production, 277*, 123078.

Vinnari, V. (2024). Operational economic and environmental reporting in mining industry information management system.

Yadav, V., Gahlot, P., Rathi, R., Yadav, G., Kumar, A., & Kaswan, M. S. (2021). Integral measures and framework for green lean six sigma implementation in manufacturing environment. *International Journal of Sustainable Engineering, 14*(6), 1319-1331. https://doi.org/10.1080/19397038.2021.1970855

Yadav, V., Kaswan, M. S., Gahlot, P., Duhan, R. K., Garza-Reyes, J. A., Rathi, R.,...Yadav, G. (2023). Green Lean Six Sigma for sustainability improvement: a systematic review and future research agenda. *International Journal of Lean Six Sigma, 14*(4), 759–790. https://doi.org/10.1108/ijlss-06-2022-0132